The Search for the Gene

Bruce Wallace

THE SEARCH
FOR THE GENE

Cornell University Press

ITHACA AND LONDON

First published 1992 by Cornell University Press.

Library of Congress Cataloging-in-Publication Data
Wallace, Bruce, 1920–
    The search for the gene / Bruce Wallace.
        p.   cm.
    Includes bibliographical references and index.
    ISBN 0-8014-2680-4 (alk. paper). — ISBN 0-8014-9967-4 (pbk. : alk. paper)
    1. Genetics.   I. Title.
QH430.W344   1992
575.1—dc20   92-4174

Printed in the United States of America

♾ The paper in this book meets the minimum requirements of the
American National Standard for Information Sciences—Permanence
of Paper for Printed Library Materials, ANSI Z39.48-1984.

Contents

Preface

Upon my arrival at Cornell University in 1958, I was greeted by LaMont Cole, an ecologist and colleague whom I had met several years earlier at one of Cold Spring Harbor's famous symposia. His words of welcome contained a message: Genetics is dead. You should be with the ecologists; the other geneticists should be in biochemistry. Nothing remains any longer of what was originally genetics.

The Search for the Gene has been written in part to counter Cole's assertion of thirty-odd years ago. Genetics, from the outset, has had a special place in my heart. As a freshman at Columbia College in 1937—one of three zoology majors that year; all others in the beginning zoology course were junior and senior premedical students—I was advised almost immediately to purchase a dictionary of Greek and Latin terms. Indeed, much of zoology in that era consisted of complex terminology—not only anatomical structures and names of organisms but also explanations! Events were explained by means of words. Explanations of evolutionary change remain most firmly in my mind: orthogenesis, autonomous development, aristogenesis, orthoevolution, hologenesis, and nomogenesis among others. Words! Little wonder, then, that I welcomed Mendel's experiments and their interpretation when I was exposed to them in my sophomore year, and Alfred H. Sturtevant and Theodosius Dobzhansky's reconstruction of phylogenies in the genus *Drosophila* through the analysis of overlapping inversions as revealed by the giant chromosomes of salivary gland cell nuclei. Genetics stressed the interpretation of observations, the construction of theoretical models (not necessarily mathematical ones), and the subjection of these models to novel tests. Classical biology, if I can use that term without being pejorative, relied heavily on verbal descriptions of observations that, when articulated with sufficient skill, became accepted as explanations. The most articulate proponent became, in fact, the most prominent one as well.

Nevertheless, as more and more ancillary sciences have availed themselves of genetic knowledge, the flavor of genetics as a science has changed greatly. On the one hand, biochemists (and their intellectual offspring, molecular biologists) have become increasingly concerned with the physical chemistry and biophysics of macromolecules; much of this concern is needed to perfect the application of what are essentially engineering processes to matters of health, agriculture, and industry. On the other hand, those involved with ecological, behavioral, or evolutionary problems often postulate a gene for this or that trait and, having postulated it, investigate its future in a hypothetical population (by computer as a rule). The calculations yield answers, of course, whether such a gene exists, could exist, or does not exist.

The Search for the Gene is my attempt to bring the logic of genetics into focus once more. "How we have gotten where we are" is often displaced from textbooks to make space for "where we are" and "what we can do with what we know." Even those who are not students but who read newspapers and newsmagazines or watch science documentaries on television know where we are. Without being told, students and nonstudents alike may come to accept "where we are" as a given—that "where we are" has always been. Thus, knowledge would join kinds, or species, as part of divine creation: on the eighth day, the Lord instilled in us knowledge concerning the structure of DNA. Such an impression would be wrong, and to some extent even dangerous. Just as we as human beings have reached our present physical state by a long process of biological evolution, so has our present knowledge of molecular biology been arrived at by a series of strenuous intellectual exercises. In the text that follows, I emphasize the intellectual aspect of genetics and follow its development, one shaky step at a time. Occasionally, some technical information must be understood in order to comprehend the intellectual advance of that moment; each item of a technical nature has been presented in a separate essay. The main text, as far as I am able, has been allowed to flow freely.

A further concession to ease of reading is reflected in the virtual absence of footnotes and subscripts. This is not a book intended to impress scholars.

References to individual publications are made in the standard way: the author's name is followed by a year (and page, if necessary); full citations are assembled at the end of the book. However, I wish to acknowledge the enormous help that I have obtained from particular publications; certain authors—in addition to noting specific citations to their work—may have a sense of déjà vu as they encounter paraphrased passages. These authors, and their works, are marked by asterisks in the list of references. I thank them and acknowledge my debt to them. I also acknowledge the comments made by colleagues J. O. Falkinham, Khidir Hilu, Stephen Scheckler, Franklin Stahl, Thomas Starling, and Bruce J. Turner, as well as those contributed by two anonymous reviewers. And I thank Sue Rasmussen for carefully and patiently altering the typescript as it evolved.

BRUCE WALLACE

Blacksburg, Virginia

The Search for the Gene

1 Like Begets Like . . .

The search for the gene began thousands of years ago, certainly before the practice of rearing animals for sacrificial slaughter or the origin of agriculture. The persons involved have long since perished; consequently, our surmises about their observations and conclusions must be based largely on the practices and knowledge exhibited by primitive peoples today.

Like mountain climbers, we might take solace, perhaps, in gazing at our goal before setting out on the nondescript lowland trail that will eventually lead us to dizzying heights. What is that snow-capped peak looming there in the distance? No, not that one; it is too close! The farther one that projects far above the tree line is our goal. Fortunately, molecular biologists have provided us with a terminus for our search. By September 30, 2005, according to Nobel laureate James D. Watson, the human genome will have been mapped. That is, all 4,000,000,000 (4×10^9) base pairs that constitute the DNA of the 23 human chromosomes will be known and will have been written down in sequence from the first base pair of chromosome 1 to the last base pair of chromosome 23. What an encyclopedia that sequence will make! A book that happens to be close at hand has 65 letters per line and 36 lines per page—or 2340 letters per page. Slightly over 425 pages of this sort would contain 1,000,000 letters; thus, 4000 volumes would be required to list the base pairs of the entire human genome. How many homes (or even schools) boast libraries of this size?

I claim that this 4000-volume library represents a logical terminus to the search for the gene not because it will be immediately understood but because all the information required to understand the hereditary informa-

1

tion needed to construct a human being will be there. Unfortunately, information not needed will be there as well. Like a computer disk that has been used over and over by the owner of a word processor—used for writing personal and business letters, for preparing articles in many drafts, and for preparing memoranda and reminders of appointments—the human genome contains currently unused segments of DNA, segments last used perhaps during the era dominated by dinosaurs, when our ancestors were inconspicuous mouse-sized mammals.

So much for the terminus of our search. How do we recognize the nondescript lowland trail upon which early human beings set off in their attempt to understand the variety of life that surrounded them? "Early human beings" must include even those forms whose intellects were only slightly more advanced than those of today's great apes. To use art as an example, we cannot consider the animal paintings in the caves at Lascaux (painted about 15,000 B.C.) the works of "early human beings"; these paintings were executed by exceptional artists and they rank with the works of the best painters of recent centuries (Dobzhansky and Boesiger, 1983, p. 129). Art of early human beings, if we are to look for crude beginnings, consists of purposeful incisions and scratches on the teeth and horns of reindeer; this art dates from 35,000 or more B.C.—20,000 years before what must be acknowledged to be prehistoric excellence.

Before recognizing that living things reproduce themselves on a reasonably fine scale, primitive beings must have noticed differences between living things on a gross scale and sensed that these forms do in fact reproduce. Conscious understanding must be included in these specifications—an understanding that can be taught to one's offspring, for example—otherwise we get involved with the ability of birds, rodents, and other mammals to choose appropriate items of food from an enormous natural menu. Ernst Mayr (1963, p. 17) reports: "Some 30 years ago [about 1933] I spent several months with a tribe of superb woodsmen and hunters in the Arfak Mountains of New Guinea. They had 136 different vernacular names for the 137 species of birds that occurred in the area, confusing only two species. It is not, of course, pure coincidence that these primitive woodsmen arrive at the same conclusion as the museum taxonomists, but an indication that both groups of observers deal with the same, non-arbitrary discontinuities of nature." The recognition of these discontinuities, the realization that generations of individuals replace one another through time, and the ability to instruct others in such recognition and realization were all essential to the eventual understanding of heredity.

In his book *Crops and Man,* Jack Harlan (1975, p. 26) describes the manufacture of arrow poison by Pygmies. "The process [is] long, complex, and dangerous for the poisons [are] extremely potent. Ingredients of 10 different plants [are] used; 8 [are] poisonous and 2 [are] gums to be impregnated with poison and stuck to the arrowheads. Two animal poisons [are] also included: beetle larvae and venum [*sic*] of a horned viper." He continues by saying that the Pygmy poison maker performs a delicate, dangerous, and highly skilled sequence of operations as exacting as some modern professions. Clearly, the art of making poison-tipped arrows is an ancient one. Modern Pygmies are as skilled in their profession as were the artists at Lascaux 15,000 years ago (i.e, skilled even by today's standards); one must go back tens of thousands of years to encounter tribes that were still experimenting with and getting to know the useful (and often dangerous) properties possessed by the plants and animals of their environment.

Understanding the succession of generations in the case of mammals is not difficult: mother cats bear kittens, cows have calves, and mares have foals. The new generation literally emerges from the old. Even in the case of birds and reptiles, the connection between mother, egg, and chick is not obscure. Upon finding an ostrich's nest, a Bushman can mark it as his private property; other Bushmen then leave it alone under penalty of death. There can be no doubt in the Bushman's mind as to either the source of the eggs in the nest or the source of baby ostriches.

Plants present much more complex life histories than do most animals. In many instances the seeds are minuscule and are dispersed at great distances from the parent plant. Even as late as World War II, British botanists were uncertain as to the source of plants that quickly appeared in bomb craters in the city of London. Were previously buried seeds now germinating because of their exposure to air, rain, and light? Or were seeds simply blowing into the new craters and taking root there? The evidence favored the latter view inasmuch as the species of plants found growing in the craters were notorious opportunistic dispersers.

Despite the sometimes perplexing life histories of many plants (life histories that students in botany and genetics courses are often required to learn), primitive peoples know them quite well. Harlan (1975, p. 23) describes a number of instances in which hunter-gatherers plant seeds or otherwise practice primitive agriculture. Indians in Nevada, for example, burn the vegetation in the fall, and in the following spring they sow the seeds of several wild plants on the burned area without tilling the soil. Certain Hindus, having dug yams, are obliged to replace the tops in order to fool the goddess Puluga, who owns all yams, cicada larvae, and beeswax. In fooling the goddess, of course, the natives are also replanting yams for the following year. Finally, the Australian aborigines honor a law stating that no plant bearing seeds can be dug up after it has flowered. Thus, it appears that human beings living in widely separated areas of the earth exhibit an understanding of the reproductive cycles of plants and exercise care (for a variety of motives) to see that the current harvest will not be the last.

The phrase "like begets like," the title of this chapter, suggests that prehistoric human beings were clever enough to notice the inheritance patterns of particularly valuable or intriguing aspects of useful wild plants or animals. That is, these primitive people knew not only that elk give birth to elk or that yam tops produce yams for harvesting but also that individual variants tend to produce offspring possessing the same unusual characteristics. Assyrians and other ancient palm growers propagated their date groves by means of suckers, not seeds. Suckers reproduce the parental plant exactly, including its sex. Plants produced by seeds are either male or female; only the latter produce dates. Except for the pollen they produce, male plants are an unnecessary waste; pollen from one male palm tree can fertilize dozens of females. Much earlier, striking family resemblances may have been discerned among related members of individual tribes.

J. B. S. Haldane (1939, pp. 187–189) disputes the view of conscious selection of nondomesticated varieties by early farmers: "What has actually happened, I think, is this. As you domesticate a plant such as wheat, as soon as you start gathering the seed and sowing it you will *automatically* select the most fertile plants" (emphasis added). The key to Haldane's argument is *evolution,* the obvious tendency for living organisms of all sorts to change in morphology or behavior over time in response to challenges posed by their environments. Before Charles Darwin's enormous treatise *On the Origin of Species by Means of Natural Selection* appeared in 1859, most people explained such changes by citing the willfulness of living beings; that is, individuals willed their evolving characteristics. Either that, or organs that were used grew in size while those that were idle dwindled. Without understanding the mechanism of inheritance, Darwin nevertheless hit upon the proper explanation for evolutionary change: individuals differ from one another. To some extent, these differences are hereditary. In any given environment, some variants survive and reproduce more successfully than others. Hence, if the environment changes, the heritable characteristics of the population undergo comparable changes as the result of the poor survival and reproduction of some individuals in contrast to the better survival and reproduction of others. A population, that is, takes on the characteristics of its successfully reproducing members; otherwise, it becomes extinct.

The automatic selection Haldane referred to results in the evolution of the domesticated crop. This principle, he explains, "only applies to plants in which we actually use for our purposes those parts [e.g., seed grain] with which it reproduces" (p. 188). Suppose, he continues, you wish to buy seed for meadow or orchard grass; you may, for example, want to raise several sheep. The seed you buy from the ordinary seedsman will produce meadow grass that is markedly poorer in quality than that which you could raise by planting wild seed. Why? Because the seedsman will have collected wild seed and then planted it repeatedly in order to obtain sufficient seed to sell. In doing so, he or she will automatically have selected plants that produce more and more seed but at the expense of the growth of other parts of the plant such as its leaves. In this instance, however, you want plants that produce heavy crops of leaves for your sheep to graze on. Obtaining commercial grass seed that produces healthy grass suitable for grazing requires very careful selection indeed. Only the seed from large and productive plants can be collected each generation for raising the next one; most seed merchants find the extra effort needed for this pattern of selection too costly.

The domestication of plants leads to a series of predictable changes, changes so well known that they are referred to collectively as "the syndrome of domestication" (Table 1-1). Wild perennial plants, for example, produce a variety of seeds; some germinate immediately, while others germinate only after one or more years have elapsed. The delay in germination may be caused by chemical inhibitors that gradually leach from the seed, or germination may require the softening of the seed coat by molds. Under a pattern of domestication that involves the harvesting of seed grain during the fall following a spring planting, only the immediate germinators contribute to next year's seed. All delayed-germination seeds are lost. Consequently, the domesticated plants become uniformly rapid germinators within very few generations—four generations in an experiment involving various species of finger millet (Hilu and deWet, 1980).

Another characteristic of wild plants (peas this time, although the same phenomenon is true of many cereal grasses) can be mentioned in conjunc-

Table 1-1. The syndrome of domestication that results automatically from planting the previous year's harvested cereal seed. (After Harlan, 1975, p. 127.)

Increase in the proportion of seed recovered
Specific adaptations:
Nonshattering
Fewer but larger flowers*
Larger seeds
Uniform ripening

Increase in seed production
Specific adaptations:
Increase in percentage seed set
Otherwise sterile flowers become fertile
Increase in flower size
Increase in number of flowers*

Increase in seedling competition
Specific adaptations:
Larger seed size
More carbohydrate, less protein
Loss of germination inhibitors
Reduction in extraneous layers on seeds

*Corn, sorghum, and millet have evolved toward few but very large inflorescences (flowers); wheat, barley, and rice have evolved toward many more inflorescences. The difference arises from the initial anatomy of the plant involved.

tion with domestication. In this case, I have had personal experience. I once collected all the seedpods from each of nine wild sweet pea plants. Each pod, properly labeled, was placed in a small vial for drying. My ultimate intention was to weigh each pea in every pod and analyze the variation in weight among peas in a pod, among pods on a plant, and among the nine plants.

The first pod I attempted to study was never included in the final analysis! As I drew this pod from its vial there was a sharp "pop"; peas from that pod literally bounced off every wall of the large room in which I was sitting. Wild peas do *not* merely drop their seeds on the ground near the parental plants. On the contrary, the pod is constructed so that as it dries it builds up considerable tension, which is relieved only when the sutures along the length of the pod tear and the sides of the pod coil up like steel springs. The dried peas are thrown for distances of ten or more yards. Domesticated peas do not possess pods that spring open in this traditional wild-type manner; only peas from pods that could be quietly opened by hand were available to early farmers for planting. Natural selection (i.e, unwitting selection) under domestication quickly eliminated the spring mechanism of wild pea pods. The morphological differences in the pods of domesticated and wild peas are so striking that pods found at archaeological sites 8000 years old can be identified as "domesticated."

Early farmers tended their crops skillfully and carefully. Grain that comes in heads (millet) or on ears (corn) requires counterintuitive behavior on the part of the farmer for successful agriculture: the best heads or ears must be saved for planting, not eating. This may seem to be a simple choice, but those who practice it are behaving in an extremely sophisticated way. All primitive peoples today set aside their best heads of millet or ears of corn for the next year's crop. When individual grains are pooled, the most fertile plants will, of course, automatically be correspondingly over-represented in the grain that is set aside for seed.

A second example illustrating the acute powers of observation that characterized early farmers can be found among the wheats; similar examples could be cited for many other cereal grains that feed the major portion of the world's population. Wheat, the grain that supplies us with the bulk of the flour used in making bread, cakes, macaroni, and biscuits, consists of a number of species. Some of these (*Triticum monococcum,* for example) have 7 pairs of chromosomes; others (*T. dicoccum* and *T. durum*) have 14 pairs; still others (*T. spelta* and *T. vulgare*) have 21 pairs.

These different wheats serve different purposes because the flours made from them have different characteristics. Thus, a 21-chromosome-pair wheat (*T. aestivum*) is a hard-kernel wheat whose flour is used in making bread and fine cakes. Flour from *T. durum* (14 chromosome pairs) is used in making macaroni, spaghetti, and related foods. Soft wheat flour (also *T. aestivum,* but a different variety) is used for piecrust and biscuits.

These different sorts of wheats have not arisen merely from evolutionary changes that occurred automatically following domestication. On the contrary, the 21-chromosome-pair species occurred first by the accidental hybridization of a domesticated wheat with a nearby wild relative; there are no 21-chromosome-pair wild species. The origin of a new highly polyploid species required that a single odd grain (or odd plant) be noticed by an early farmer who then saved it and planted it. The sequence would have involved the accumulation of sets of 7 similar (but not identical) chromosomes with each hybridization: the 7-paired species would have produced 14-paired ones through the "natural" hybridization of wild species; any one of these by hybridizing with a 7-chromosome species could have produced a 21-

paired species. In each of the latter instances, someone had to notice an abnormal kernel (or plant), save it, plant it, and then determine whether the flour obtained from the new type was useful and tasteful. The outcome has been the evolution of wheat species through shrewd observation and subsequent selection; some of the early species that originated in Turkey and the Near East are now of little or no commercial value. Their germ plasms may, of course, be useful to wheat breeders for disease or drought resistance or for characteristics currently unappreciated.

A similar account could be presented for barley, another cereal grass. Wild barley heads have two rows of seeds; "two-rowed" barley is found in ancient archaeological sites and is grown even today in adverse (arid) environments. Most commercial barleys, however, are "six rowed"—they have six rows of grain per head. Barley of this type arose under domestication and was saved by early farmers. There are no six-rowed wild species of barley, nor do six-rowed varieties persist for long in nature after escaping from cultivation.

Before concluding this chapter, I should say something about the ideas that primitive peoples held concerning the transmission of hereditary traits. One such view is illustrated in Figure 1-1. Central American women, the farmers of the family, often raised different varieties of corn for different purposes (much as rural Americans today have a variety of apple trees: some for pies, others for apple sauce, and still others for eating raw). Experience revealed that these varieties could contaminate one another. Consequently, when the women of a village laid out their garden plots, they planted the varieties (labeled A and B here) in the pattern shown in the diagram. All A plantings were clustered and placed at a considerable distance from the B plantings, which were also clustered. These women knew nothing about pollen, but the clustering of plantings on different family garden plots increased the frequency of successful pollinations. Whenever an A kernel appeared on a B ear, or vice versa, the interpretation was that a root had grown beneath the garden and hence that the contamination was by underground rather than by airborne means.

The book of Genesis (chapter 30) presents a novel explanation for the inheritance of coat color in sheep and goats. Jacob, having carefully tended his father-in-law's herds so that they increased in size, now wished to return to his own country. However, he was prevailed upon by his reluctant and wily father-in-law (and uncle), Laban, to remain and tend the herds for still another year. In lieu of wages, he struck the following bargain:

Figure 1-1. Native gardeners, each with a small personal plot, arrange their plantings of two varieties of corn so that the harvest of one strain tends not to be contaminated by cross-fertilization by the other. The four intersecting corners of adjacent plots are planted with the same variety. In addition to reducing intervarietal contamination, this planting pattern enhances the overall seed set in each of the four plots by concentrating the pollen in one area.

Thou shalt not give me anything: if thou wilt do this thing for me, I will again feed and keep thy flock.

I will pass through all thy flock today, removing from thence all the speckled and spotted cattle, and all the brown cattle among the sheep, and the spotted and speckled among the goats: and of such shall be my hire.

So shall my righteousness answer for me in time to come, when it shall come for my hire before thy face: every one that is not speckled and spotted among the goats, and brown among the sheep, that should be counted stolen with me. (Gen. 30:31–33)

Laban shrewdly removed from the herds all ring-streaked and spotted goats as well as those that had any white in their coats, and all the brown sheep; these he gave to his sons. This culling left Jacob with only animals that he would be unable to retain as his own. Furthermore, Laban separated the two flocks—Jacob's and his own (and his sons')—by three days' travel, thus preventing any interbreeding.

Undaunted, Jacob "took him rods of green poplar, and of the hazel and chestnut tree; and pilled white streaks in them, and made the white appear which was in the rods." These spotted and streaked rods were then placed around the watering troughs where mating animals could see them. Brown lambs were set aside for Laban. Weak animals were also set aside for the father-in-law, but the strong animals were forced to mate in the presence of the streaked rods. According to the biblical account, "the man [Jacob] increased exceedingly, and had much cattle and maidservants, and menservants, and camels, and asses."

Agriculture: An idea or a family tradition?

Clearly, early views regarding biological inheritance were molded by "a need to know"—that is, a need to know who and what one's neighbors are in the biological community. Hunters and gatherers needed this information; the great advance these primitive peoples made over the rest of the animal kingdom lay in their ability to remember and transmit vital information from one individual to another by word of mouth or other cultural means. (Daws and crows use the same technique. In any flock of these birds, there is usually at least one who recognizes any dangerous object or being and who, during a confrontation, informs all others by an appropriate warning call. Thereafter, all who have seen the danger and have heard the accompanying warning call are able to teach other, naive, birds in turn.)

Agriculture greatly increased the need to know about inheritance and inheritance patterns. Although many of the changes undergone by domesticated plants and animals may merely represent the syndrome of domestication, many others reflect the skill exhibited by individual farmers who chose seed or vegetative cuttings with considerable care.

While the question does not bear directly on the search for the gene, we might ask whether agriculture as a way of life was an *idea* or a *set of procedures* perfected and spread by its practitioners. Was it, for example, an idea such as Catholicism, which was spread worldwide largely by the preachings of celibate priests? Or was it a trade whose practice was limited largely to certain peoples who increased in number, extended their homelands, and brought the practice with them?

A solution to the above question was sought by three scientists—P. Menozzi, A. Piazza, and L. Cavalli-Sforza (1978)—through a comparison of the archaeological evidence on the spread of agriculture from the Near East into Europe (an event that started 9000 and ended about 5000 years ago) with data concerning the frequencies of different genes in the various regions of Europe. If agriculture spread as an idea and if the hunter-gatherer populations of Europe were genetically different from the farmers of the Near East, that difference should remain. The conversion of a Polynesian village to Catholicism by a missionary priest, to cite an analogy, has no effect on the genetic composition of the local population.

If, on the other hand, farmers extended their homelands from the Near East and Turkey into Greece and the Crimea and eventually through central Europe into Germany and France, one might expect that at each step along the way (the rate of spread has been estimated at 1 km per year) the farmers and their hunter-gatherer neighbors intermarried. The result would be a gradient of genetic variation extending from genotypes characteristic of peoples inhabiting the Near East to those characteristic of the early peoples of central Europe. A gradient of precisely that sort was found by the three scientists; the gradient, like the spread of agriculture itself, extends from the southeast to the northwest of Europe (see Figures 1-2 and 1-3). Thus, the spread of agriculture from one of its several centers of origin (the Near East) into Europe accompanied the movement of human populations; it was not merely the dissemination of a good idea. Indeed, if one can judge by the resistance of today's hunter-gatherers to agriculture (Harlan, 1975, pp. 48–

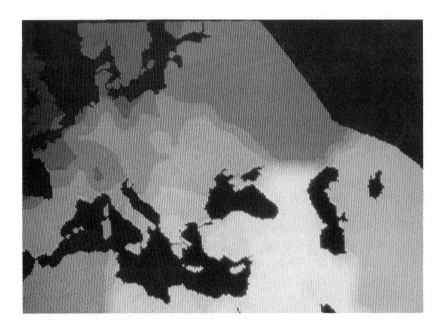

Figure 1-2. The spread of early farming from the Middle East into Europe. The lightest cross-hatching denotes areas where farming was present earlier than 8500 B.P. ("before the present"); the seven other gradations of cross-hatching (lighter to darker) denote areas where farming was introduced in 8000–8500 B.P., 7500–8000 B.P., 7000–7500 B.P., 6500–7000 B.P., 6000–6500 B.P., 5500–6000 B.P., or later than 5500 B.P. (From Menozzi et al., 1978, copyright 1978 by the AAAS.)

49), farming does not appear to be an excellent idea to everyone. Farmers both need (for labor) and can support larger populations than can hunter-gatherers, who must keep their numbers considerably smaller because they dare not overharvest the untended plant and animal communities upon which they rely for food. Thus, as agriculture spilled over from the Near East into the neighboring areas of Europe, the farmers probably outnumbered the nearby tribes of hunters. The intermarriage of the (invading) many with the (resident) few produced the genetic gradient that exists today.

This account can be terminated by reference to an analogous modern example involving sorghum and the native tribes of Africa. Different tribes—by accident, by preference or tradition, or as the result of stresses imposed on sorghum plants by their local environments—have developed recognizably different varieties of sorghum. In areas where members of

Figure 1-3. A contour map, comparable to the one in Figure 1-2, based on the frequencies of 38 alleles at nine gene loci in human beings. The more dissimilar the shade, the more dissimilar the genetic composition of the populations involved. Hence, the map reveals a genetic gradient, or *cline*, extending from the Middle East (where agriculture was first practiced) into and across Europe. The more recently agriculture has arrived in a given region, the greater the populations of that region deviate genetically from populations inhabiting the Middle East. (From Menozzi et al., 1978, copyright 1978 by the AAAS.)

two tribes meet and intermarry, the varieties of sorghum also intergrade through hybridization (K. W. Hilu, personal communication; see Stemler et al., 1975).

Box 1-1. Genetic distance

The curiosity of some readers may be aroused when I speak of two genetically quite distinct groups—persons living near the North Sea and others living in the Near East, for example—and then mention a gradient of populations, a *cline* in proper terminology, that joins the two.

How does one speak of distances when speaking of the genetic composition of populations? Easily! Many techniques are used to measure the genetic distances between populations, but the calculation first used for this purpose is still valid, especially for purposes of illustration.

Consider a right triangle:

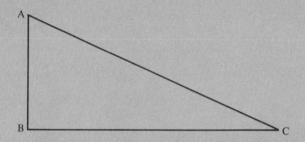

What is the distance from A to C (a distance that can be represented as $\overline{AC}$)? From high school algebra we know that $\overline{AC}^2 = \overline{AB}^2 + \overline{BC}^2$, or $\overline{AC} = \sqrt{\overline{AB}^2 + \overline{BC}^2}$.

Consider now a room or box such as the following:

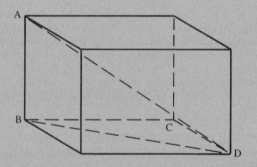

What is the distance $\overline{AD}$? By examining the lines shown in the diagram we can see that $\overline{AD}^2 = \overline{AB}^2 + \overline{BD}^2$. We can also see that $\overline{BD}^2 = \overline{BC}^2 + \overline{CD}^2$. Hence, $\overline{AD}^2 = \overline{AB}^2 + \overline{BC}^2 + \overline{CD}^2$. That is, the length of the diagonal from an upper corner of a room to the lower opposite corner ($\overline{AD}$) equals the square root of the sums of the squares of the three dimensions of the room, $\sqrt{\overline{AB}^2 + \overline{BC}^2 + \overline{CD}^2}$.

Pictures are limited of necessity to three dimensions, but if we notice the change from the two-dimensional example to the three-dimensional one, we can infer how to incorporate additional dimensions into the calculations:

$$z^2 = a^2 + b^2 + c^2 + d^2 + e^2 + \ldots$$

Suppose that the frequencies of two alleles at each of four gene loci (a, b, c, and d) in each of four populations are as follows. (Note that the sum of the frequencies of each pair of alleles [a_1 and a_2, for example] equals 1.00.)

Population	a_1	a_2	b_1	b_2	c_1	c_2	d_1	d_2
1	.90	.10	.80	.20	.50	.50	.30	.70
2	.30	.70	.60	.40	.10	.90	.40	.60
3	.40	.60	.50	.50	.30	.70	.50	.50
4	.10	.90	.30	.70	.20	.80	.80	.20

The distance squared (d^2) between populations 1 and 2 is calculated as .36 (that is, $[.90 - .30]^2$) + .36 + .04 + .04 + .16 + .16 + .01 + .01 = 1.14; $d_{1\cdot2} = \sqrt{1.14} = 1.07$. The distance squared ($d^2$) between populations 1 and 3 is calculated as .25 + .25 + .09 + .09 + .04 + .04 + .04 + .04 = 0.84; $d_{1\cdot3} = 0.92$.

In a similar manner, the following distances can be computed:

$$d_{1\cdot4} = 1.57$$
$$d_{2\cdot3} = 0.37$$
$$d_{2\cdot4} = 0.77$$
$$d_{3\cdot4} = 0.68$$

These results reveal that population 1 is quite different from the other three ($d_{1\cdot2} = 1.07$, $d_{1\cdot3} = 0.92$, $d_{1\cdot4} = 1.57$). Populations 2 and 3, on the other hand, are quite similar ($d_{2\cdot3} = 0.37$). Population 4 is quite different from populations 2 and 3 ($d_{2\cdot4} = 0.77$, $d_{3\cdot4} = 0.68$) but not as different as it is from population 1 ($d_{1\cdot4} = 1.57$). Thus, one can draw the following diagram:

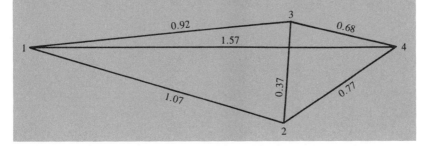

The relationships are not linear, only approximately so. Population 2 is farther from both populations 1 and 4 than is population 3. Thus, the four populations are best represented (only approximately at that) as occupying a surface.

Although the procedures used by Menozzi et al. (1978) are considerably more complex than the one used here for illustrative purposes, the outcome was similar: genetic distances between dozens of populations were calculated, and these were found to form a systematic surface pattern that slopes from the Near East and southeastern Europe to areas in the northwestern portions of the European continent. The contours of this surface closely match those of a surface that had been developed independently by archaeologists studying the spread of agriculture from the Near East into the hunting-gathering tribes of Europe. Hence, the conclusion that early farmers spread their genes (by intermarriage with local inhabitants) as well as their grains and their agricultural know-how appears to be justified.

2 Ages of Reason

In the late 1500s, Montaigne, a famous French essayist, wrote, "I am in conflict with the worst, the most mortal, and the most irremediable of all diseases—kidney stones" ("Of the Resemblance of Children to Fathers," 1580). He then continued, wondering how his father, who had died of kidney stones at age 75, had transmitted the disease to his son (the only one of three children similarly affected), when the son (Montaigne) had been born twenty-five years before his father's first symptoms. "And he being then so far from the infirmity, how could that small part of his substance wherewith he made me, carry away so great an impression for its share. And how so concealed, that till five and forty years after, I did not begin to be sensible of it?" Montaigne's puzzlement was such that he concluded by saying, "He that can satisfy me in this point [by providing a reasonable explanation], I will believe him in as many other miracles as he pleases." There was, however, one condition that Montaigne imposed for credibility: the explanation could not be more complex than the problem itself. Montaigne's caveat illustrates the essence of scientific inquiry and explanation: a presumptive explanation should not raise more (and more difficult) problems than the problem it attempts to solve.

Insight into the type of explanation that Montaigne would not accept can be gained by considering the views regarding reproduction, sex determination, and variation held by scientists of the sixteenth century. Still under the influence of Aristotle and the physicians Hippocrates and Galen, these scientists thought that semen originated from surplus food or from particles secreted by all body parts. Semen was either formed in the brain or

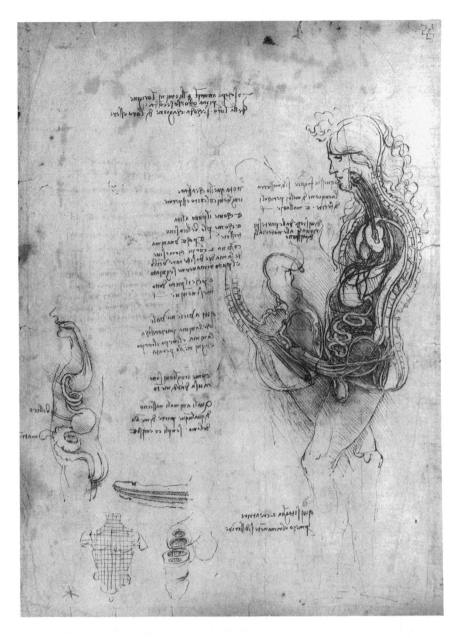

Figure 2-1. Leonardo da Vinci's pen-and-ink drawing of the coition of a hemisected man and woman. The tube connecting the penis to the sacral area of the spinal cord illustrates the prevailing belief that the thickest part of semen, that which carries the soul to the future embryo, comes from the spinal cord. (Leonardo da Vinci, *Coition of Hemisected Man and Woman*, R.L 19097v. Windsor Castle, Royal Library. © 1991 Her Majesty Queen Elizabeth II.)

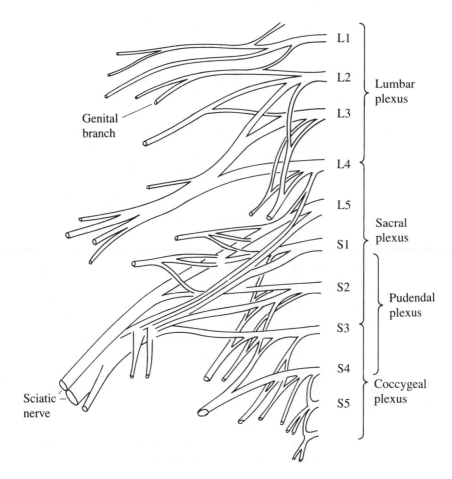

Figure 2-1 (cont.). The true complexity of these lumbar, sacral, pudenal, and coccygeal nerve plexi. Although the penis is innervated by minor nerves (the genital branch arising from nerves L1 and L2) from this region of the spinal cord, a firm belief in the prevailing (and erroneous) view was needed to represent the anatomy as da Vinci did. "Seeing" homunculi in human sperm was an error of the same sort—one sees what one believes should be seen.

deposited there by the blood, after which it was transferred to the penis by way of the spinal cord (Figure 2-1).

Leonardo da Vinci, noting that children of black parents living in Europe were black, whereas a child born of a white woman but fathered by a black man was gray, argued that the seed of the mother is equal in strength to that of the father. Despite this and similar observations, a centuries-long controversy eventually arose between spermists and ovists, both preformationists but bitterly divided as to which germ cell contained the tiny preformed individual.

Montaigne might have profited by studying the Jewish law on circumcision, which was available during his era. In 1565 the *Code of Jewish Law,* compiled by Rabbi Joseph Karo, was published; in it was the following paragraph: "If a woman lost two sons presumably from the effects of the circumcision, as it was apparent that their constitutions were so weak that the circumcision has caused their exhaustion, her third son should not be circumcised until he had grown up and his constitution became strong. If a woman lost the child because of the circumcision *and the same thing happened to her sister* [emphasis added], then the children of the other sisters should not be circumcised until they have grown up and have a strong constitution."

During the thousands of years that Jews practiced circumcision, their religious leaders had detected the pattern with which this operation proved to be fatal to newborn male babies. Someone had noticed the familial aspect of these calamities. Knowledge of such cases presumably had to accumulate through story telling (oral history). Only then could one or more thoughtful persons discern the patterns and sense the probabilities with which circumcised babies died (of hemophilia).

Probability and pattern were the two aspects of heredity stressed considerably later by Pierre Maupertuis, a disciple of Newton, a geographer, and later the head of the Berlin Academy of Sciences (see Glass, 1947). After arriving in Berlin (ca. 1740), Maupertuis sought for and found a family (the Ruhes) in which some members possessed extra fingers and toes (polydactyly). The pedigree he assembled is shown in Figure 2-2. The polydactylous woman at the start of this pedigree is Elisabeth Horstmann; Elisabeth Ruhen, her daughter, was the mother of Jacob Ruhe, the man initially brought to Maupertuis's attention. That gentleman had two polydactylous sons as well as several affected siblings whose sex is not reported.

In his effort to understand the concentration of polydactyly in this family, Maupertuis attempted to estimate the frequency of this disorder among the inhabitants of a city of 100,000 persons. He found two afflicted men. If he had overlooked another three (which he doubted), the frequency of persons with extra fingers and toes would still then be only 1/20,000. That an af-

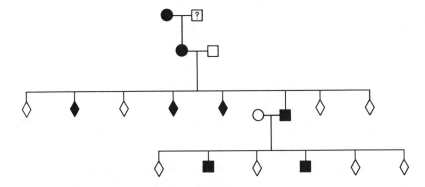

Figure 2-2. Polydactyly in the Ruhe family according to the pedigree investigated by Maupertuis, who concluded that the concentration of so rare a trait in one family over three generations could not be a matter of chance. The question mark indicates that no information is available for the great-grandfather in this pedigree. Lines that connect male (□) and female (○) symbols indicate marriages. Offspring—male, female, and of unknown sex (◇)—are connected to their parents by vertical lines that eventually connect to marriage lines. Solid symbols indicate the presence of polydactyly.

fected person should have an affected child, and that child have an affected child in turn, would, over the course of three generations, have a probability equal to $1/20{,}000 \times 1/20{,}000 \times 1/20{,}000$, or $1/8{,}000{,}000{,}000{,}000{,}000$. Maupertuis compared his conclusion concerning the inheritance of polydactyly to the certainty that physicists assign to certain physical phenomena such as the daily reappearance of the sun, which has risen fewer than 2 trillion times.

Maupertuis reached an additional conclusion. With no firm knowledge of the physical structure or composition of the male and female "semens," he suggested that both contain particles: male particles in the one, female in the other. Many animals have litters consisting of numerous young, and thus, Maupertuis thought, the semens contain many more particles than are needed for the formation of any one embryo. For each embryo, however, he suggested that single particles of one sex match single particles of the other, thus generating thousands of paired particles.

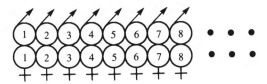

Arrays of perfectly paired particles give rise to normal children; particles that fail to retain partners or allow extra particles to intrude as triplets lead to abnormal or monstrous embryos. Why Maupertuis settled on the idea of pairs is not clear; perhaps because it represents the simplest scheme that allows individuals of both sexes to contribute to their offspring. Perhaps (and this is interesting) because even a casual examination of his pedigree suggests that although half of each family was polydactylous, the other half was normal. This point may not have been lost on a scholar who had studied under Isaac Newton. It seems not to have occurred to Maupertuis that the thousands of male and female particles may have been prepackaged and that the pairing resulted from the union of two packages or that multiple embryos reflected the presence of many packages.

The microscopists

In 1694, Rudolf Camerarius, a German botanist specializing in the sexuality of plants, wrote: "To solve this difficult question [i.e., the fate of pollen once it meets the stigma] it would be most desirable to hear from those whose optical instruments have given them the eyes of a lynx what the pollen grains from the anthers contain, how far they penetrate into the female structures, whether they remain intact until they reach the place where the seed is received, and what comes out of them when they burst" (quoted in Stubbe, 1972, p. 94). Who would imagine that fifty years after these remarkably precise statements of needed information were written, a professor of botany, J. G. Siegesbeck of St. Petersburg, shocked that there were so many pollen grains and so few seed chambers, could still write, "What man will ever believe that God Almighty should have introduced such confusion, or rather such shameful whoredom, for the propagation of the reign of plants. Who will instruct young students in such a voluptuous system without scandal?" (quoted in Olby, 1985, p. 3).

Camerarius was correct in calling for the help of microscopists; with their newly developed instruments they were discovering a universe of minute beings never seen before. Only twenty years before Camerarius asked their help in extending his observations, Antonie van Leeuwenhoek had described the red blood cells of mammals, amphibians, and fish. In 1677, Leeuwenhoek had described spermatozoa. Like Niklaas Hartsoeker

Figure 2-3. The human sperm according to Niklaas Hartsoeker, a seventeenth-century microscopist. Supposedly, within the sperm's head was a small infant with a disproportionately large head (notice the nose tucked between the knees). Only with the aid of its mother could the small infant within the sperm escape its prison. Note that although this observation proved to be wrong, it was—at that time—positive evidence. After all, if there wasn't a small person in each sperm, where else could babies come from?

(see Figure 2-3), Leeuwenhoek assumed that a spermatozoon was a preformed organism—infinitely small, though complete in all organs and limbs. The most extreme view of preformism was held by Charles Bonnet, who discovered parthenogenesis in plant lice. Bonnet contended that the first created member of a species contained all subsequent descendants preformed, each generation within the previous one, until the end of the world.

Spectacular as the observations of early microscopists were, optical equipment required nearly two centuries of improvement before decisive observations were made that truly aided in the search for the gene. The ambiguities of observations made with merely "lynx-eyed" devices led to controversies rather than to well-established conclusions. Marcello Malpighi, for example, observed defined structures in egg yolk viewed through his microscope. These early stages of embryonic development (recall that a hen's egg is fertilized in the oviduct before the shell is laid down) convinced Malpighi that the new individual was present and preformed in the egg; it became visible as its preformed parts grew in size. Thus, the ovists arose in opposition to spermists, although both groups believed that development consisted only of the enlargement of tiny preformed parts.

Walther Flemming's studies should be considered next, and at some length. He owes his exalted place among cytologists largely to two rather diverse developments. One of these was the dye industry that developed in Germany. Flemming used a variety of stains (dyes) in preparing his material, including carmine, hematoxylin, and aniline. Only gradually did the chemistry of dyes and the chemistry of cellular material develop to the point where it was understood why any one dye attached to and consequently stained a particular cellular structure. In part, especially in the case of chromosomes, the chemical interaction was related to the unique chemical structure of chromosomal material.

In addition to the chemical dye industry, the field of optics in Germany was making tremendous strides. As a consequence, by 1878 Flemming had at his disposal oil-immersion lenses capable of 1000-fold magnification. Shortly thereafter, he had access to apochromatic microscope objectives. These lenses removed the distracting halo of rainbow colors that surrounds every object of interest seen through more primitive lens systems.

Two of Flemming's illustrations are reproduced here to show the detail the new stains and lenses made visible to a keen observer. (The observer cannot be omitted from the system; Calvin Bridges, a cytogeneticist of whom I shall speak later, had extremely fine visual resolution. Unfortunately, many excellent cytologists lose their skill as they age, and the debris floating within the interior of their eyes interferes with their vision.) Both illustrations are presented because, excellent as the detail in Figure 2-4 may be, it is incapable of revealing the sequence with which the observed patterns occur. Recall that each drawing represents a cell killed by exposure to acetic acid or some other fixative, then cut into extremely thin slices (10 micrometers, or 1/100th of a millimeter), and finally exposed to dyes that differentially stain the chromosomes and the cytoplasm of the cell. Although such "snapshots" may be arranged in a logical sequence, the direction in which the sequence should be read is not revealed. The sequence is revealed by the events shown in Figure 2-5: the division of a single cell. In this case, the sequence of events (but not the detail) can be followed visually. The same sequence can then be superimposed upon the sequence of excellent snapshots obtained from fixed and stained cells.

An enormous step along the trail we are following in our search for the gene was made by Wilhelm Roux in 1883, one year after Flemming published the illustrations discussed above. After referring to the work of Flemming and other contemporary cytologists, Roux emphasized what appeared to him to be the unnecessarily complex means by which the nucleus divides; in his opinion, the "dance of the chromosomes," as it was sometimes known, demanded an explanation. *Why* should the division of a cell's nucleus be any more complex than the division of the cell itself? Cells divide either by developing a constriction at the midline that, in a relatively short time, cuts the cell in two (animal cells) or by laying down in the cell's middle a plate that spreads laterally until it reaches the cell wall, thus dividing the original cell into two smaller ones that then grow in size until they, too, lay down dividing plates (plant cells). Why couldn't the nucleus divide by a simple midline constriction?

A careful examination of the figures drawn by Flemming and others showed Roux what can be easily seen in Figure 2-4(41). As the chromosomal material condenses early in the nuclear division cycle, it can be seen to consist of double structures: point by point along the length of each chromosome, each point has been duplicated. Therefore, argued Roux, each small particle of which the chromosome is built is needed for the life of the cell. The elaborate "dance" ("rich play of forms"; see Grene and Burian, 1991, p. 434) is required for the precise division of nuclear material—point by point—because the roles of these small particles of chromatin differ; merely dividing the nuclear material into two approximately equal portions does not ensure the presence of a representative of each small particle.

Whys are not profitable questions in physics and chemistry. In those sciences, events happen or do not happen according to the physical properties of molecules and atoms. In biology, *whys* may lead to what are regarded derisively as "just-so stories." Why is the rabbit's tail so short? Why are there no feathers on the turkey buzzard's head? Rudyard Kipling and Joel Chandler Harris (via Uncle Remus) have supplied colorful answers to these and numerous similar questions. With proper care, however, biologists can—as Roux did—ask Why? The reason they can do this is because evolutionary change consists largely of an accumulation of information. This accumulation is not directed; it is based on trial and error, with

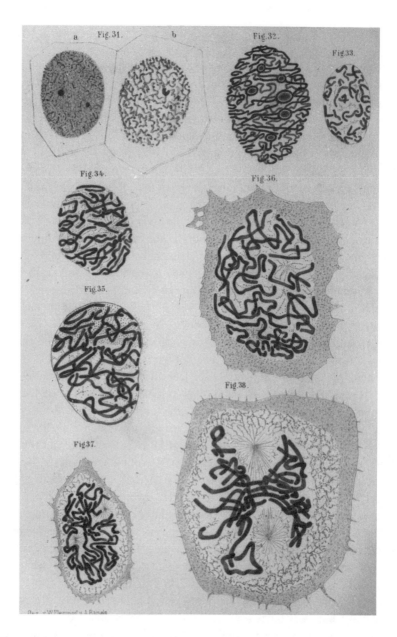

Figure 2-4. Flemming's drawings of mitosis in fixed and stained cells of a salamander embryo. Because this figure has been reproduced from Flemming's original plates, it bears his figure references (31–46) as well; these are utilized here and in the text. (31a and b) Early and late resting stages. (32–37) Early through late prophase. (38) Early metaphase. (39–41) Various views of metaphase. (42–44) Anaphase. (45) Telophase. (46) Cell division.

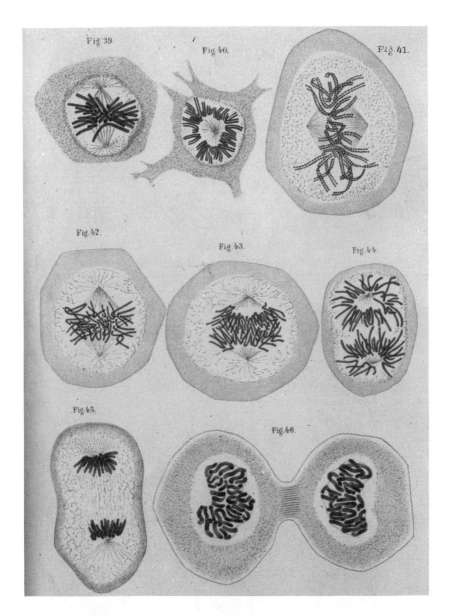

Fig 39.

Fig 40.

Fig. 41.

Fig 42.

Fig 43.

Fig 44.

Fig 45.

Fig 46.

sexual or asexual reproduction providing the means for amplifying (in numbers) the individuals who possess appropriate information. "Nothing in biology makes sense except in the light of evolution" is a phrase comparable in power to "like begets like." The "making of sense" allows biologists to ask Why? Roux exhibited considerable insight in noting that nuclear divisions are much more complex than they seemingly need to be.

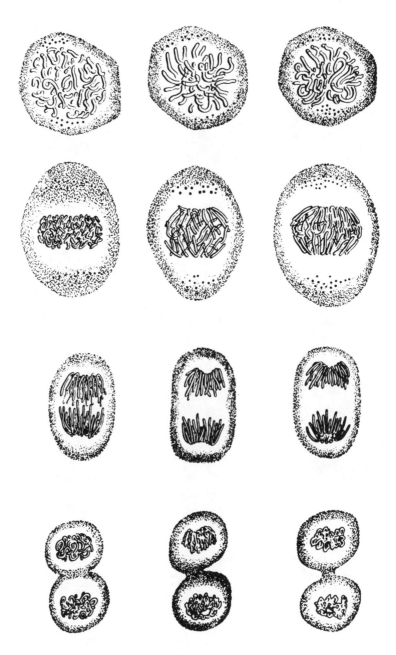

Figure 2-5. Flemming's drawings of mitosis in living epidermal cells of a salamander larva. Although living cells do not show the details that fixing and staining make visible, they provide the movement necessary to understand how one nuclear configuration leads into the next. Without these observations on living cells, the interpretation of the fixed and stained material would remain uncertain.

Two decades after Roux's surmise, proof of his conclusions was provided by Theodor Boveri in a most clever way. To appreciate the problem, examine Figure 2-4(44) once more. Do not take my count seriously, but I estimate from Figure 2-4(44) that 24 chromosomes are moving toward each pole of the division figure.* Look carefully at these chromosomes. Is there any indication that one of these darkly staining bodies has any function that differs from the others? Of course not! They all stain alike and therefore look alike.

Boveri's analysis began with an aberration. Normally, one egg is fertilized by one sperm; that fact was established for animal reproduction by Oskar Hertwig in 1875. August Gärtner had made a similar point in 1837 by showing that two or more parental types (pollen) never blend when mixed pollen is used to fertilize a flower: each seed is fertilized by a single pollen grain. If, however, eggs of an animal such as a sea urchin are exposed to a concentrated sperm suspension, some eggs are fertilized by two sperm. (Normally, when a sperm enters an egg, an alteration of the egg's membrane rapidly spreads around the surface of the egg, preventing the entrance of additional sperm cells. With abnormally high concentrations of sperm cells, a second sperm may enter the egg before the membrane has been altered at the second point of entry.) These dispermic eggs give rise to two distinct classes of developing embryos: In some, the spindle along which the chromosomes separate has three poles rather than the normal two that arise after fertilization by one sperm. In others, however, the spindle has four poles. These alternative outcomes of dispermy can be represented as follows:

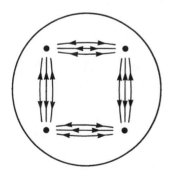 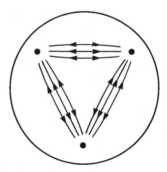

*Since writing this I have checked an atlas of chromosome numbers for various species of amphibians and reptiles; Old World salamanders—numerous genera and species—have a diploid number of 24, just as I estimated from this century-old figure.

In brief, Boveri noted that 58 of 719 embryos arising from three-pole dispermic eggs developed normally; however, no embryos of approximately 1200 four-pole dispermic eggs underwent normal development. He then reasoned as follows: Each of the sea urchin's 18 chromosomes (36 chromosomes is the normal diploid number) has a function that is necessary for the development of an embryo and differs from that of all other chromosomes. If so, normal development may ensue if *at least one* of each of the 18 chromosomes is present in every cell of an embryo.

Using Boveri's specified rules of the "game"—that is, the conditions under which he predicted an embryo would survive—we can make our own calculations (Box 2-2). The outcomes of these calculations are the following: (1) about 12% of all embryos that develop from three-pole eggs should survive (Boveri found that 8% of 719 embryos survived), and (2) only 3 in 100,000 embryos developing from four-pole eggs should survive (Boveri saw none— "kein einziger"—in nearly 1200 embryos examined).

Boveri's analysis is one of the classic studies in developmental biology. By careful reasoning, he accounted for the contrasting results observed regarding the fate of abnormal embryos, those developing from dispermic eggs. E. B. Wilson, one of the world's leading cytologists of that era, wrote in 1918: "One who, like the writer, had puzzled in vain over the riddle presented by double-fertilized eggs of sea urchins could not read Boveri's complete and beautiful solution without a thrill; and it may be doubted whether a finer example of experimental, analytical and constructive work, compressed within such narrow limits—the paper on multipolar mitosis comprises but twenty pages and is without figures—can be found in the literature of modern biology" (p. 75). Such is the respect that one scientist pays to another for completing an exceptionally beautiful analysis. "Thrill" is not an extravagant term to use in describing one's reaction upon encountering such an analysis.

Where, one may ask, is the advance over Wilhelm Roux's conclusion of 1883? Roux suggested that every *particle* on every chromosome plays a different and essential role in the life of a cell; Boveri (1902) showed that every *chromosome* plays a different and essential role. Roux's inference arose from an attempt to understand why the division of cell nuclei is so much more complicated than the division of cells themselves; his conclusion was based on patterns of chromosomal duplication provided by contemporary cytologists. Boveri's conclusions rested on the mathematical analysis of the segregation patterns of chromosomes on multipolar spindles. The analysis predicted that three- and four-polar spindles should lead to strikingly different outcomes, and his prediction was confirmed by direct observation of nearly 2000 abnormal embryos. There is, I believe, even today a schism that separates scientists (such as theoretical physicists) who are comfortable with conclusions reached by reason alone and those who insist (as experimental physicists do) on performing tests on material substances. Neither type has a monopoly on Truth; both theoretical reasoning and experimentation have led people to ridiculous conclusions. Truth is approached, however, by the alternate application of both methods of research: intellectual effort plus careful experimentation.

Formation of germ cells

I introduce this topic not to make the reader relive the times in high school and college biology during which he or she committed the processes of mitosis and meiosis to memory, but rather to emphasize still one more inference made by early cytologists. Karl Wilhelm von Nägeli, an outstanding nineteenth-century biologist, had foreseen the problem created by the union of parental idioplasms (nuclear contents) with each fertilization: the amount of such material in cells would double in each generation. Hertwig (1890) posed the problem more succinctly by asking, in effect, How is a summation of idioplasm avoided in successive generations? He answered his own question by saying that paternal and maternal idioplasms are divided into two equal parts just before fertilization. Substituting "number of chromosomes" for "idioplasm," the answer might read, "The doubling of the number of chromosomes that would occur following the fusion of two nuclei during fertilization is prevented by an opposing process known as reduction." And, for the moment, this is the aspect of meiosis that I wish to stress. During the formation of germ cells in virtually every species of plants and animals, maternal and paternal chromosomes (with some exchange of parts) are separated without replication during one meiotic division. Hence, germ cells come to contain only one-half as many chromosomes as normal somatic (or body) cells; the full (diploid) number is restored when the pronuclei of the egg and the fertilizing sperm fuse.

I also cite a paragraph that appeared in *An Atlas of the Fertilization and Karyokinesis of the Ovum,* by E. B. Wilson (1895):

These facts justify the conclusion that the nuclei of the two germ-cells are in a morphological sense precisely equivalent, and that they lend strong support to Hertwig's identification of the nucleus as the bearer of hereditary qualities. The precise equivalence of the chromosomes contributed by the two sexes is a physical correlative of the fact that the two sexes play, on the whole, equal parts in hereditary transmission, and it seems to show that the chromosomal substance, the *chromatin*, is to be regarded as the physical basis of inheritance. Now, chromatin is known to be closely similar to, if not identical with, a substance known as *nuclein* ($C_{29}H_{49}N_9P_3O_{22}$, according to Miescher), which analysis shows to be a tolerably definite chemical composed of nucleic acid (a complex organic acid rich in phosphorus) and albumin. And thus we reach the remarkable conclusion that *inheritance may, perhaps, be effected by the physical transmission of a particular chemical compound from parent to offspring.* (Emphasis added)

Wilson later changed his mind regarding the chemical basis of heredity (he switched to protein rather than nucleic acid); nevertheless, five years before the rediscovery of Mendel's classic paper, the "home" for heredity and for familial similarities was being located in chromosomes, physical bodies whose importance for life had been (or, in Boveri's case, was about to be) demonstrated, not only as intact bodies but also (Roux's earlier paper) piece by piece.

The plant hybridizers

The distinction may not withstand careful scrutiny, but I like to view scientists as either those who are comfortable only if they are working with actual things (biochemists, physiologists, embryologists, and cytologists) or those who are at ease studying patterns and from them inferring underlying causes (geneticists are among these). The weaknesses of both this classification and reasoning by inferences are illustrated by the remarks of Thomas Hunt Morgan (1909), then an embryologist but soon to become a geneticist, made at a convention of animal breeders: "In the modern interpretation of Mendelism, facts are being transformed into factors at a rapid rate. If one factor will not explain the facts, then two are invoked; if two prove insufficient, three will sometimes work out. The superior jugglery sometimes necessary to account for results may blind us, if taken too naively, to the common-place that the results are often so excellently 'explained' because the explanation was invented to explain them. We work backwards from the facts to the factors, and then, presto! explain the facts by the very factors that we invented to account for them."

I cite Morgan's comment because I want to go back in time and join a second branch of the trail followed by some whom we now recognize as being in search of the gene—the botanists who intercrossed different varieties and species (definitions that clearly separated these terms had not yet been enunciated, even today some confusion remains) and noted the characteristics of their hybrid offspring. At the start, following the publication of Linnaeus's *Systema naturae,* hybridization experiments were carried out in light of Linnaeus's claim that "species are as numerous as there were created different forms in the beginning." Consequently, the rationale for hybridization experiments evolved slowly, and testing the creative powers of God remained a valid reason for some time. Having observed fertile, true-breeding hybrids, Linnaeus concluded that God had created monospecific genera and that the species currently existing had arisen through crossbreeding, presumably among locally differentiated varieties. Although he believed that species had *not* been created and fixed for all time, Linnaeus (according to Stubbe, 1972, p. 99) still did not concede evolutionary theory as we understand it today— in fact, he rejected such ideas.

Josef Kölreuter, an extremely energetic hybridizer during the mid-1700s, convinced himself that hybrid plants (he worked initially with species of tobacco) were intermediate between their two parents, thus contradicting the expectations of the preformationists. "Two different homogeneous liquid substances," he claimed, "are necessary for the procreation of every plant. These two substances, which are ordained by the Creator of all things to unite with one another, are male and female seed" (quoted in Stubbe, 1972, p. 102). He went on to compare a hybrid plant to the salt produced by mixing an acid and a base.

The questions scientists pose for themselves determine the nature of the experiments they perform and the conclusions they reach on the basis of their observations. A. D. Hershey, one of my colleagues at the Biological Laboratory (see Box 7-3) during the 1950s, received the ultimate in praise from a French virologist who wrote, "You have answered questions that I had not even thought to ask." During the late 1700s, questions about creation and the nature of species dominated much of the research done by plant hybridizers. Thomas Knight, a British horticulturist, was interested in increasing agricultural yield through hybridization (either by taking advantage of hybrid vigor or by creating new combinations of agriculturally useful traits). In experiments with carefully emasculated peas which were then crossed with other varieties, Knight frequently observed luxuriant growth in the hybrids. He noted that the gray color of seeds is dominant to white. He also noted that backcrosses of the hybrid with its white-seeded parent gave rise to plants bearing both gray and white seeds; he did not, however, count these seeds or attempt to calculate a ratio.

John Goss, another British horticulturist, also chose peas for his experimental material. Crosses of blue-seeded (female) and white-seeded (male) plants produced only white seeds. These white seeds produced plants whose pods contained all white, all blue, or both kinds of seeds. Furthermore, when planted in an isolated plot, the blue seeds produced plants whose pods contained only blue seeds. Yet white seeds produced plants whose pods contained only white seeds as well as plants whose pods contained peas of both colors.

Thomas Laxton, writing in 1872, said: "I have noticed that a cross between a round white and a blue wrinkled pea, will in the third and fourth generations . . . at times bring forth blue round, blue wrinkled, white round, and white wrinkled peas in the same pod, that the white round seeds when again sown, will produce only white round seeds, that the white wrinkled seeds will, up to the fourth or fifth generation, produce both blue and white wrinkled and round peas, that the blue round peas will produce blue wrinkled and round peas, but that the blue wrinkled peas will bear only blue and wrinkled seeds" (quoted in Stubbe, 1972, p. 110). As was the case with both Knight and Goss, Laxton neglected to count the numbers of each type of plant (or seed).

I stressed the absence of numerical data as if that lack prevented the horticulturists from deducing the pattern of inheritance in peas. Let me cite a counterexample involving one of biology's greatest thinkers. Charles Darwin was intrigued by numerous scientific problems. The origin of species, for example. Or, the origin of oceanic atolls. He carried out a tremendous study of the effects of self- and cross-fertilization in plants; among the species he studied was *Zea mays*. Darwin's was the first study of hybrid vigor in corn. He was also interested in heredity; many of his experimental results are presented in his book *The Variation of Animals and Plants under Domestication* (1875). In these studies, Darwin concentrated his attention on the transmission and prepotencies of appearances; he did not (as Mendel had done) consider the physical entities responsible for appearances. Hence, he failed to appreciate his numerical data.

Among Darwin's plants were snapdragons. He reported (1875, 2:46) that a cross between the common (asymmetric) snapdragon and one with peloric (i.e., unusually regular) flowers produced no peloric hybrids in two large beds. However, a cross of normal hybrids with normal hybrids produced among 127 plants, 90 that were normal (i.e., asymmetric) and 37 that were peloric. In this case, unlike those above, the numerical results were presented—but they remained uninterpreted.

Learned societies of past centuries frequently posed questions and sought essays that provided answers; the essays were then judged and their authors were eligible for cash awards. Thus, in 1819, the Berlin Academy of Sciences posed the question: "Is hybrid fertilization possible in the vegetable kingdom?" In 1830, the Dutch Academy in Haarlem changed the question so that it related more to practical matters of plant culture. (In his response to this invitation, August Gärtner made the following important point: if a flower is fertilized with a mixture of pollen [its own and pollen from a foreign source], the male characteristics are not blended in the offspring. Hence, each egg appears to be fertilized by but one pollen grain.)

The Paris Academy of Sciences announced an open competition in 1861 on the topic "plant hybrids from the point of view of their fertility and the permanence or impermanence of their characters" (Stubbe, 1972, p. 120). The two persons who responded to this challenge, D. A. Godron and Charles Naudin, both concentrated on peripheral matters. Godron wondered whether the characters of hybrids reproducing by self-fertilization remained constant or if hybrids that are perpetuated by selfing eventually revert to original parental forms; this was important when the basic question concerned patterns of creation. Naudin was more concerned with interspecific than with intraspecific hybrids. Even so, he drew conclusions that were extremely insightful, including the observation that it is in the formation of pollen and eggs that elements representing species' essences are most clearly separated one from the other.

At this point, the search for the gene (made largely by people who did not recognize the exact goal of their search—that is, of their research) can be summarized. Proceeding backward, we have seen that physical characteristics of parental strains are passed from generation to generation. Some traits are *dominant*—a descriptive term that means only that each may

conceal an alternative trait, one that may reappear in a subsequent generation. Physical characteristics (*phenotypic* traits) can assort themselves into
combinations not present in parental varieties. More often than not, numerical data and ratios were missing from early reports; the example provided
by Darwin was a notable—but unutilized—exception.

Cytologists armed with new dyes and improved lenses were making
advances. The nuclei of eggs and sperm were recognized as being equivalent, thus paralleling Gärtner's observation that hybrid individuals are in
many—but not all—respects a nearly perfect average of their parents.
Boveri demonstrated that each chromosome (of the sea urchin, at least) has
a unique and vital function with respect to the survival of embryos. Boveri
proved on a gross scale (the chromosomal level) what Roux had inferred
nearly twenty years earlier about the individual chromatin particles that
constitute dividing chromosomes.

And then, nearly lost in antiquity, was Maupertuis's suggestion that male
and female particles pair, particle by particle, and that the entire array
of such pairs—with no absentees (deficiencies) and no duplications—is
needed for normal development; duplications and deficiencies lead to
monstrous individuals. Maupertuis had no knowledge regarding cells in
general or spermotozoa and eggs in particular. In his view, two fluids—
seminal and menstrual—mixed and gave rise to embryos. Multiple births,
including rather large litters in some species, "proved" that more particles
are present in these fluids than are necessary for a single embryo. How,
then, did Maupertuis settle on *pairs* as being the required number? I
suspect, with no proof, that the equal number of normal and polydactylous
individuals in his pedigree influenced his thinking. The equal numbers of
normal and hemophiliac sons of seemingly normal mothers could have had
a similar influence on ancient rabbis, but they were not searching for the
gene—they were concerned with rules that would keep a religious rite from
proving fatal to newborn male babies.

Box 2-1. Human pedigrees

Human geneticists have devised a shorthand representation of family
structure: squares represent males, circles represent females, open figures signify that the individuals are normal (or at least do not exhibit the
trait under discussion), and solid figures represent individuals that ex-

hibit the trait of interest. If the sex of an individual is unknown, a diamond-shaped symbol may be used to indicate this uncertainty. Lines connect husband and wife; others lead to their children.

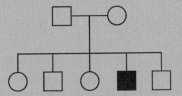

The Jewish law concerning circumcision can be diagrammed as follows. The solid dots within seemingly normal females reflect the basis for both the pattern and the probabilities:

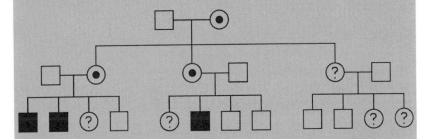

The first point ("If a woman lost two sons . . .") is illustrated at the left. Two sons are hemophiliacs who died following circumcision; all sons subsequently born are thus exempt from the operation until grown to manhood. (Hemophiliac sons under the environment prevailing in biblical times would probably have died of wounds and bruises accidentally inflicted upon them as children.) The solid dot shown in the mother's symbol indicates that she may transmit the fatal condition to sons as yet unborn.

The second point ("If a woman lost a child . . . and the same thing happened to her sister") is revealed by the three related families. It now becomes possible to insert a solid dot into the grandmother's symbol because the law suggests that all her daughters may have hemophiliac sons. The increased probability that the untested daughter may even yet produce hemophiliac sons is implicit in the exemption of her sons from circumcision.

This Judaic law does not explain the physical basis of hemophilia (and in that sense would not respond to Montaigne's [p. 17] question), but it does reveal an understanding of two prerequisites for answering that ultimate question. A pattern of inheritance and a recognition that observed events are not independent of one another (i.e., chance events) must be understood before the deeper question about causation can be answered.

Box 2-2. Boveri's analysis

In defending the calculations of early population geneticists, J. B. S. Haldane (1964) approvingly stressed the precision with which verbal statements must be expressed if they are to be fitted to a mathematical model. Otherwise, as he pointed out, inconsistent statements can be assembled in the guise of a logical argument (I am reminded here of the old song that contains the doubly self-contradictory phrase "a barefoot boy with shoes on stood sitting on the grass"). Although I am generally inclined toward Haldane's position, I have argued (1988) that verbal arguments need not be imprecise; they can be expressed nonmathematically with sufficient precision to make appropriate experimental tests possible. Mathematics, that is, has no monopoly on precision of thought, and indeed, the simplifying assumptions often invoked to render a problem amenable to mathematic analysis frequently lead to caricatures of life rather than to acceptable solutions for real-life problems.

With that preamble, let us examine Boveri's hunch and the experimental data he gathered in its support. Many would use the term *hypothesis* rather than *hunch*, but I believe they can often be used interchangeably: a hypothesis is a formalized hunch. Boveri believed that although all the chromosomes of an organism (the sea urchin, for example) may look alike when examined under the microscope (especially after they have been fixed and stained), each plays a different and indispensable role in the life of cells. Consequently, he proposed that in order to live and function, a sea urchin cell must have at least one representative of each of its 18 chromosomes.

He set out to test this proposition by studying the fate of sea urchin eggs that had been fertilized by two, rather than the usual single, sperm. These dispermic eggs fell into two classes: those that developed three-poled spindles for the initial chromosomal segregation (and hence three primary cells), and those that developed four-poled spindles (and correspondingly four primary cells). These two types of division patterns differ greatly in the proportion of viable embryos they produce. Boveri calculated the difference that might have arisen if at least 1 chromosome of each of the 18 different ones in the haploid set were essential for development. Recall that a dispermic egg would have three haploid sets: one contributed by the egg and one by each of the two sperm cells.

Consider the three-pole spindle:

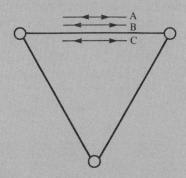

All three representatives (A, B, and C) of the first of the 18 chromosomes might become oriented on one spindle as shown; if so, the third pole (which eventually corresponds to one of the three embryonic cells) will not receive a representative of that chromosome. Any other configuration will lead to at least one representative of the first of the 18 chromosomes in each of the three cells. Now, the probability that all three representatives (A, B, and C) will lie on one spindle equals 1 (i.e., the first representative, A, must go to one of the 3 spindles) $\times \frac{1}{3} \times \frac{1}{3}$, or $\frac{1}{9}$. The probability that all three will *not* go to one spindle, then, equals $1 - \frac{1}{9}$, or $\frac{8}{9}$. Furthermore, the probability that each of the 18 chromosomes will be oriented so that at least one representative of each will be found in each of the three cells equals $(\frac{8}{9})^{18}$, or 0.1200. In a study of 719 three-polar eggs, 86.3 would be expected to survive; Boveri found that 58 of 719 such embryos survived.

Now, consider the four-pole spindle:

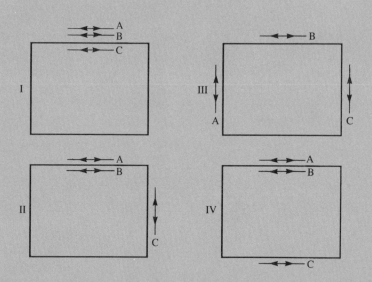

The possibilities in the case of a four-poled spindle are much greater than in the case of the three-poled one.

As shown in I, all three representatives of any one of the 18 chromosomes might come to lie on a single spindle. The probability equals 1 (recall that the first representative must go to one spindle of the four) × ¼ × ¼, or ¹⁄₁₆.

Another possibility, shown in II, has two representatives lying in one spindle and the third in an adjacent one. The odds in this case are 1 × ¼ × ²⁄₄ = ²⁄₁₆ (the second representative joins the first, and the third goes to either of the two adjacent spindles) and, another possibility, 1 × ²⁄₄ × ²⁄₄ = ⁴⁄₁₆ (the second representative goes to either of two spindles that are adjacent to the one first occupied; the third representative then joins either the first or the second). The total probability for this configuration equals ⁶⁄₁₆. The patterns represented in diagrams I and II lead to primary cells that lack this chromosome; these cells are inviable, and the developing embryo does not survive.

If diagrams I and II depict patterns that have an overall probability of ⁷⁄₁₆ (i.e., ¹⁄₁₆ + ⁶⁄₁₆), then the next two patterns (diagrams III and IV) must have a combined probability of ⁹⁄₁₆. This can be confirmed by noting these two possibilities.

One pattern (diagram III) has one representative of a given chromo-some in each of three adjoining spindles. This pattern can arise in two ways. The probability for the first equals 1 (i.e., the first representative, A, must go to one of the four spindles) $\times$ ¾ (the second goes to either of the two adjacent spindles) $\times$ ¾ (the third goes to either of the two still-vacant spindles) = ⁴⁄₁₆. The second pattern leads to a probability that can be calculated as 1 $\times$ ¼ (representative B goes to the spindle that is opposite the first) $\times$ ¾ (representative C goes to either of the two still-unoccupied spindles) = ²⁄₁₆. The overall probability of this pattern equals ⁴⁄₁₆ + ²⁄₁₆ = ⁶⁄₁₆.

The pattern represented in diagram IV can also arise in two ways with the following probabilities: (1) 1 $\times$ ¼ (B goes to same spindle as A) $\times$ ¼ (C goes to the spindle lying on the opposite side of the four-poled figure) = ¹⁄₁₆; and (2) 1 $\times$ ¼ (B goes to the spindle lying on the opposite side of the four-poled figure from (A) $\times$ ¾ (C joins either A or B) = ²⁄₁₆. Thus, the configuration shown in diagram IV has a total probability of occur-ring equal to ¹⁄₁₆ + ²⁄₁₆ = ³⁄₁₆. The total probability of patterns III and IV is ⁶⁄₁₆ + ³⁄₁₆ = ⁹⁄₁₆, as expected.

Diagrams III and IV represent configurations that lead to the presence of at least one representative (A, B, or C) of the given chromosome in each of the four primary embryonic cells that arise from a four-poled spindle. The probability that all 18 chromosomes will have a representa-tive in each of these four cells equals $(9/16)^{18}$, or 0.000032. This is a very small probability, and indeed, Boveri saw no viable embryo among the nearly 1200 he studied. Thus he concluded that a cell, in order to survive, must have at least one representative of each of the 18 chromo-somes included within its nucleus. Expressed differently, each of the 18 chromosomes performs a function essential for life, one that cannot be performed by any of the remaining 17. As a footnote to this analysis, I might add that Boveri's "kein einziger" has in some English translations mistakenly become "one" rather than "not a single one."

Box 2-3. Microscopy and histochemistry

Examination of once-living material requires that it be immobilized, that different sorts of materials be identified, and that small objects be magnified for examination. Take, for instance, the study of a hen's egg. One might crack it and empty the contents of the shell into a bowl; that procedure reveals a yolk surrounded by a bowl-shaped watery egg white. Alternatively, one can boil the egg for seven or eight minutes, remove the shell, and cut the now-hard-boiled egg with an egg slicer. No effort is required to identify yolk and egg white; both are solid, and each has a characteristic texture and color. Whether a hard-boiled egg is sliced lengthwise or on a bias, crosswise, the result is a circular segment of yolk surrounded by solidified white albumen. Consequently, it appears that egg yolk is suspended within the egg albumen as a sphere.

Living—or at least raw—tissue must be "fixed" before it can be cut into thin slices. Freezing is adequate if slices are to be removed by a razor blade. For much thinner slices, the tissue is fixed by chemicals (formaldehyde is a favorite) that cross-link protein molecules, thus rendering them insoluble. The more rapidly the fixative penetrates the tissue, the less distorted are the final results.

Following fixation, a time-tested process exists for replacing the water that occurs in all living tissue by alcohol, then by xylol, and then by paraffin—all this without seriously affecting the network of now-cross-linked protein molecules. The tissue, thoroughly embedded in paraffin, is then sliced much like boiled ham but on an extremely delicate slicer called a microtome. The individual slices may be as thin as 5 or 6 micrometers (0.005–0.006 mm).

After attaching such sliced material onto a glass slide (egg albumen is the glue), the paraffin can be dissolved and replaced once again by water. Now the tissue can be exposed to stains or dyes. As an accident of history, Flemming had available for his studies stains that were the result of the newly founded German organic dye industry. Carmine and hematoxylin, two natural dyes, were already in use by cytologists in 1865. Aniline, the mainstay of the dye industry, was first synthesized in 1856, and cytologists were using aniline-based dyes (notably eosin) by the 1880s.

The interactions between dyes and cellular constituents are terribly complex chemical reactions; they are not fully understood even today. Some dye molecules react with specific amino acids, the building blocks (20 in number) of proteins. The relative proportions of the different amino acids may differ from protein to protein; hence, the colors with which different proteins stain may differ. Other dyes react with the linkages that hold amino acids together within the (linear) protein molecule;

$$\begin{array}{c} O \\ \| \\ -N-C- \\ | \\ H \end{array}$$

these reactions tend to make all protein molecules stain the same color. DNA (deoxyribonucleic acid) and RNA (ribonucleic acid), two cellular components discussed at length in later chapters, also combine with specific components of various dyes. One procedure (the Feulgen stain) utilizes a well-known organic chemical reaction to stain (and hence reveal the presence of) DNA alone. A whole branch of biology known as histochemistry has developed; its practitioners specialize in the use of chemical dye reactions for revealing ever-finer aspects of cell and tissue structure. Many of the techniques developed by histochemists have been utilized as well by persons who separate tissue components in gel matrices by electric currents and then reveal the presence of specified substances (especially enzymes) by the use of carefully selected stains. Such matters will recur at intervals as we continue our search for the gene.

Optics, which had started (more or less) with Leeuwenhoek grinding simple lenses during the seventeenth and early eighteenth centuries, became a subject of mathematical investigation that led by the mid-1800s to compound microscopes equipped with apochromatic lenses. The latter, ground from types of glass with different refractive indexes, correct the otherwise divergent paths of red and blue light (the extremes of the visual spectrum) so they converge on a single point. The result of this correction is the removal of the troublesome multicolored halo that otherwise would surround the object being viewed.

The upper diagram in Figure 2-6 illustrates the paths of light that result in the magnification of a small object by an ordinary hand lens. The light is bent in such a manner that the eye perceives the object as being farther away and larger. The lower diagram in Figure 2-6 illustrates a compound microscope. In this case, the upper lens magnifies not the small object itself but the already enlarged image of that object, thus creating a greatly enlarged virtual image that is perceived by the observer.

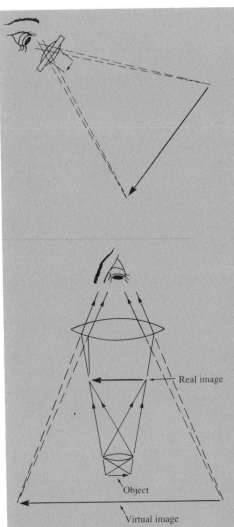

Figure 2-6. The magnification of an object by a simple lens (top) and a compound lens (bottom). In effect the eye sights along the incoming light paths (paths established by the convergence of light in the eye) and, as a consequence, the brain perceives an image (which may or may not be real) that is larger than the object itself. The wavelengths of visible light range from about 4000 Å (blue) to 6500 Å (red). Light perceived by the microscopist must be reflected from or transmitted by the object being examined; objects whose size approaches that of the wavelengths of light do not reflect light efficiently. Therefore, the magnification that can be achieved by light microscopes has theoretical, as well as practical, limitations.

The light microscope has physical limitations that result in an upper limit to magnification of about 1000-fold, a limit already reached by the latter half of the nineteenth century. The limit is based largely upon the wavelength of light; ultraviolet rays allow one to "see" (photograph, since human beings cannot see ultraviolet light) more detail than can be seen with ordinary light sources. As a practical matter, objects closer to one another than 0.1 micrometer (0.0001 mm) cannot be distinguished;

they are perceived as one object. In a later chapter I discuss DNA, a helical molecule that is 20 Å (0.000002 mm) in diameter. This diameter is about 50 times smaller than the lower limit of light microscopy; if DNA is to be made visible, a technique other than light microscopy will be required.

3 Gregor Mendel

Genetics is a unique science: it has a beginning—1900. That is the year three botanists—Karl Correns of Germany, Hugo de Vries of the Netherlands, and Gustav Tschermak of Austria—announced the results and interpretations of their investigations of hybridization and then conceded that the same interpretations had been reached by Gregor Mendel thirty-five years earlier. Mendel's results were published in 1865 in a paper now renowned for its style and the clarity of its logic, but unnoticed for decades. This paper has been recommended reading for genetics students of three generations; a few of these students may actually have taken the recommendation seriously. The vast majority, unfortunately, have learned of Mendel only through secondary sources.

Historians who have carefully investigated the rediscovery of Mendel's work have taken some of the luster from that event (see Olby, 1985). None of the three rediscoverers has escaped the suspicion that his conclusions were arrived at by applying Mendel's reasoning to his own data, although all claimed to have seen Mendel's report, which reached similar conclusions, only after having interpreted their own data correctly.

Mendel's report is a pleasure to read. I do not intend to review the data and propound on Mendel's laws of segregation and independent assortment; too many elementary biology texts do precisely that. What I do intend to do is consider some of Mendel's concerns, explain what biologists have (erroneously, I believe) come to view as his archaic (and deficient) symbolism, and emphasize in what way he outshone his predecessors.

Eight pages into his paper, following an account of his predecessors' failures and a review of the various physical traits exhibited by different species of peas (*Pisum* spp., or should they be regarded as varieties of *P. sativum?*), Mendel illustrates the variation in the proportions of seeds exhibiting dominant and recessive traits among individually tested plants. The data illustrating this variation are presented in Box 3-1. The question Mendel raises, and which prompted him to present these data, is important and recurs later in his paper when he makes the following point: "This [the distribution $A + 2Aa + a$] represents the *average* course of self-fertilization of hybrids when two differing traits are associated in them. In individual flowers and individual plants, however, the ratio in which the members of the series are formed may be subject to not insignificant deviations. Aside from the fact that the numbers in which both kinds of germinal cells occur in the ovary can be considered equal only on the average, it remains purely a matter of chance which of the two kinds of pollen fertilize each individual germinal cell."

To the extent that he is referring to the role of chance in determining the hereditary composition of individual eggs when alternative possibilities exist (*A* versus *a,* for example) or to the chance events involved with fertilization itself, Mendel is on solid ground. But if he is prepared to say that one *Aa* plant will produce one extreme ratio of germinal cells (rather than 50:50 as his laws demand) only to be counterbalanced by another plant whose germinal ratio is the reverse (i.e., an admission that his 50:50 average ratio rests on a between-plant average), Mendel is on extremely shaky ground. For this reason, I feel that it is necessary to demonstrate in Box 3-1 that the seemingly divergent ratios exhibited by Mendel's ten plants do not diverge more than one would expect on the basis of chance alone. No reason exists for believing that these plants differed in their germinal ratios. The same is probably true for the variation Mendel saw from pod to pod.

What difference does it make whether Mendel's concern regarding inter-plant variation was justified? Had he been correct, the search for the gene would have been infinitely more complex than it has proven to be. If different plants whose genetic designation (say, *Aa*) was identical proved nevertheless to exhibit enormous variation in the proportions of *A* and *a* gametes they produced, the physical basis of heredity—the gene, as we now call it—would have been ephemeral indeed. Let's merely give a sigh of relief that *Aa* plants are homogeneous in germinal ratios, and that the flowers borne by these plants are homogeneous as well.

In his writings, Mendel used both single and double symbols: *A, Aa, a,* for instance, or when discussing two characters, *A B, A Bb, Aa Bb, Aa b,* and the like. These symbols and the more common ones (introduced by William Bateson [see Punnett, 1928]—*AA, Aa,* and *aa,* for example—can be compared after the following extremely brief review of Mendel's main observations and conclusions:

1. Given two varieties of garden peas that differed in certain obvious respects, Mendel convinced himself that each one could be maintained generation after generation without exhibiting any variation; that is, these varieties were true breeding. Seven such varieties were chosen for study.

2. When two true-breeding parents were crossed, their F_1 (first filial) hybrid offspring were uniform and exhibited (in every instance) the obvious aspect of *one* variety with no hint of the other's presence; that is, one of the two traits was dominant over the other. This was true of each of the seven pairs of contrasting traits.

3. When the F_1 hybrids were self-pollinated, individuals exhibiting the recessive trait reappeared in one quarter of all progeny of that (the F_2, or second filial) generation. (This observation is noted in Box 3-1 where I mention Mendel's genius in saying $3:1$ rather than adhering to the actual data with their $2.98:1$ ratio.) Plants exhibiting the recessive character proved to be true breeding upon further testing.

4. The F_2 individuals exhibiting the dominant trait were subjected to further tests. Mendel found that one-third of these plants were true breeding like the original parent while two-thirds produced recessive offspring just as the F_1 hybrids had done. These fractions led him to the proportions 1 *A* : 2 *Aa* : 1 *a* for the F_2 progeny.

5. Having noted that a $1:2:1$ ratio would occur if both the male and the female gametes of *Aa* plants had been one-half *A* and one-half *a,* and, further, if these germ cells had combined at random,

Male \ Female	½ A	½ a
½ A	¼A	¼Aa
½ a	¼Aa	¼a

Mendel stated that the underlying physical causes of these two traits *segregate* during the formation of germ cells. (Tests of this and other, more complex, predictions were made, among others, by crossing F_1 hybrids to their true-breeding recessive parents. Here, the proportions of offspring reveal the proportions of gametes; Mendel's predictions were sustained in all tests.)

6. When following two or three pairs of alternate traits, for each of which one member of the pair was dominant and the other recessive, Mendel found that the outcomes could be fitted to distributions generated by combining $1:2:1$ ratios, that is, $(1:2:1) \times (1:2:1) = 1:2:1:2:4:2:1:2:1$, or, an even more complex case, $(1:2:1) \times (1:2:1) \times (1:2:1) = 1:2:1:2:4:2:4:8:4:2:4:2:1:2:1$. (This observation led Mendel to conclude that during the segregation of alternate characters, the alternate characters of different pairs *assort* at random; thus, not only are all possible combinations of characters obtained, as Mendel's predecessors had already shown, but also these combinations occur in predictable ratios. Test crosses involving complex hybrid plants mated with true-breeding recessives confirmed Mendel's conclusions regarding the formation of germ cells.)

A review of the summary provided at the conclusion of Chapter 2 reveals that many of the details that influenced Mendel and led to his conclusions had been observed by his predecessors. What Mendel contributed (and these contributions are critical to understanding heredity) were (1) precise numerical data, (2) an insightful interpretation of these data, (3) an understanding of the further consequences if this interpretation were correct, and (4) a confirmation of these predicted consequences.

With the preceding having been said, we can return once more to Mendel's seemingly quaint notation for true-breeding (*A* and *a* for a single trait such as round or angular seeds) and hybrid plants that fail to breed true (*Aa*). Biologists now routinely use *AA, Aa,* and *aa* to indicate that true-breeding plants also carry two "factors," or genes; the two happen to be indistinguishable, however. I think that Mendel, perhaps because of his training in physics and surely because of his facility with arithmetic (algebraic) calculations, intended to emphasize a different point—a probabilistic one: *A* and *a* individuals produce only one kind of gamete; the single letter emphasizes the 100 % . *Aa* individuals produce two kinds of gametes in equal proportions.

Thus, to determine the number of different types of gametes that an individual of genotype *A Bb C D e Ff g Hh* would produce, Mendel had only to multiply 2 (*Bb*) by 2 (*Ff*) by 2 (*Hh*), and determine that there would be eight kinds of gametes. This calculation could be made mentally while drawing one's finger down the list of symbols. No one before Mendel could make this precise statement (indeed, too few genetics students today can do so).

When two plants of differing genotypes are crossed, Mendel's symbolism leads to rapid calculations of the proportions of both phenotypes (appearances of individual plants) and genotypes (the actual combinations of genes carried by individual plants) among the progeny. For example, in the following cross

$$Aa \quad B \quad c \quad Dd \quad Ee \quad F \quad G$$

$$\times$$

$$A \quad Bb \quad C \quad d \quad E \quad Ff \quad g$$

the progeny will fall into one or the other of only two contrasting phenotypes: D or d. With respect to genotypes, the progeny are much more diverse; one can calculate that 32 different genotypes are generated by this cross ($2 \times 2 \times 1 \times 2 \times 2 \times 2 \times 1 = 32$).

To reachieve Mendel's simplicity, biologists need to insert gametes into diagrams that illustrate matings; such diagrams reveal that all (100%) of the gametes produced by an *AA* individual are *A*. Unfortunately, authors of most elementary textbooks, both college and high school, omit the gametes and consequently depict matings in the following meaningless way:

$$AA \times Aa$$

	A	*a*
A	*AA*	*Aa*
A	*AA*	*Aa*

I leave it to the reader to recognize that once the logic of 100% *A* gametes is abandoned and the two horizontal rows, each labeled *A,* are depicted instead, the diagram suggests that males and females are capable of making only two germ cells each—an obvious absurdity.

Mendel's affection for numerical calculations is evident throughout his 1865 paper. He calculates the proportion of hybrid (i.e., *Aa*) plants that would remain on an island if it were to be inhabited initially by a single self-fertilizing *Aa* individual. Within five generations only 1 individual in 32 would be such a hybrid; all other plants would be true-breeding *A* or *a* types. (Recall that a self-fertilizing *AA* or *aa* plant can produce only *AA* or *aa* progeny.) He illustrates with calculations why a holistic approach in which species of plants are distinguished by overall appearances (rather than by specified traits) leads to meaningless discussions concerning the eventual reversion of hybrid offspring to original "types" (compare with p. 35). Knowledge concerning individual characteristics that differentiate the species as well as their dominance relationships is essential for reaching a sound conclusion; general impressions can seriously mislead the observer. An account I read many years ago (but, unfortunately, can no longer cite) claimed that the emphasis on "Mendelian ratios" that has plagued students in elementary biology and genetics courses for decades was an emphasis imposed by the rediscoverers of Mendel's paper, not by Mendel himself. While it may be true that Mendel established his ratios only as an intermediate step in his own search for the gene (i.e., the material basis of heredity), his familiarity with numbers and his use of numbers in illustrating his insights could not help but impress less mathematically inclined biologists. His rediscoverers merely set out to teach their students what they themselves and their predecessors had failed to appreciate: the essential importance of recognizing patterns of inheritance and of interpreting the probabilities with which contrasting characters occur in individual progenies.

Box 3-1. The chi-square (χ^2) test

The gathering of numerical data is an essential feature in most scientific investigations; a second and often unappreciated feature is the proper interpretation of the data once they have been collected. At times intuition is enough: Mendel skirted many difficult problems by merely stating, "When the results of all experiments are summarized, the average ratio between the number of forms with the dominating trait and those with the recessive one is 2.98 : 1, *or 3 : 1*" (emphasis added). Had he attempted to find a simple explanation for a 2.98 : 1 ratio, he would have failed absolutely.

The statistical method by which one can compare empirical data with those expected on the basis of some theoretical model (or hypothesis) was devised by Karl Pearson in 1900; thus the test Mendel needed was not available to him, but, if they were familiar with the mathematical literature, it *was* available to his rediscoverers. The test in question is the chi-square test (χ^2), by which one decides how often deviations from a predicted model would be expected to occur by chance. This knowledge leads to a second decision—namely, do the observed data fit (i.e., agree with) the model or should the model be rejected?

Suppose, to take a simple case, you meet a stranger in the park at lunchtime, and in the course of becoming acquainted you become involved in a game of matching pennies. As you return to work after lunch, you note that he won 60 tosses, whereas you won only 40. Pennies, being impartial, should have led to 50 wins for the stranger and 50 for you. You are now struck by the festering thought that you have been "taken" by a clever gambler.

How can you resolve your doubts? Nothing is to be gained by citing the 60:40 outcome while ranting, "He cheated," because that would require that you cite all possible patterns by which he may have cheated: on the 3rd, 6th, 15th, 17th, 21st, and 89th tosses, perhaps. No, that approach is hopelessly complex. You then declare a null hypothesis: your opponent did not cheat (and, therefore, did not cheat on any of the 100 tosses). If this were so, he should have won 50 tosses and lost the other 50 to you. Now, a chi-square test can be employed:

	Opponent	
	Won	Lost
Observed (O)	60	40
Expected (E)	50	50
Difference ($O - E = d$)	10	-10
d^2	100	100
d^2/E	2	2

The sum of $d^2/E = \chi^2 = 4.00$ with 1 degree of freedom; the probability of seeing deviations this large or larger lies between 4% and 5%. In brief, you could expect to find deviations between observed and expected values as large or larger than this by chance fewer than once in 20 comparable coin-flipping encounters.

At this point, differences in personal philosophies arise. One school maintains that *before* carrying out these calculations, you should decide at what level you will conclude, in effect, "My opponent cheated!" If, as many routinely do, you had chosen 5% as the critical level, you would be free (nay, compelled) to utter this condemnation.

My philosophy (and that of many others) differs from the above. The chi-square test has provided evidence that must now be weighed with other factors, some of which may not have been known previously. You may, for example, be called unexpectedly as a witness in a trial at which your lunchtime opponent, if convicted, will be sentenced to prison for life as a three-time offender. Will you insist that the 60:40 pattern of wins and losses is adequate for this decision? Or you may meet your opponent once more but now at the local church, wearing his priestly garments and conducting communion. Will you denounce him publicly as a cheat and a thief? I believe that in such cases you would want a much stronger bias than 60:40; perhaps 70:30, 80:20, or even 100:0 would make you more secure.

High school students and college freshmen generally receive instruction in the use of chi-square tests in conjunction with genetic data—often Mendel's but, occasionally, their own or those of some early geneticist who experimented with fruit flies. These tests correspond precisely to the coin-flipping example cited above except that the expected ratios might be $\frac{3}{4} : \frac{1}{4}$ (the F_2 of a monohybrid cross), $\frac{1}{2} : \frac{1}{2}$ (monohybrid, test-cross generation), or $\frac{9}{16} : \frac{3}{16} : \frac{3}{16} : \frac{1}{16}$ (the F_2 of a dihybrid cross). In each case, one compares the observed (O) and expected (E) numbers, calculates the difference (d) for each class, and sums d^2/E to get a chi-square value whose degrees of freedom are one less than the number of classes generated by the model.* The probability of seeing deviations this large or larger is obtained from a table in a textbook, and a decision is made either to accept or reject the theoretical model.

In completing this essay, I examine some data presented by Mendel in the early pages of his 1865 report. The question concerns the variation in

---

*Determining the degrees of freedom for a particular analysis can be a formidable task, especially if expectations are generated by the observations themselves. On the other hand, when a theoretical model generates the number of classes and the expected numbers of observations per class (given a total number of observations), then the degrees of freedom are always one less than the number of classes. The term is apt: when the number of observations (total = 100) falling in classes 1, 2, and 3 are 15, 22, and 43, the number in the fourth class *must* be 20—one has no freedom in assigning this number.

ratios that is observed among individual plants. Picking up Mendel's account, we read:

> The distribution of traits also varies in individual plants, just as in individual pods. The first ten members of both series of experiments may serve as an illustration:

		Experiment 1			Experiment 2	
		Shape of Seeds			Coloration of Albumen	
Plant	N	Round	Angular	N	Yellow	Green
1	57	45	12	36	25	11
2	35	27	8	39	32	7
3	31	24	7	19	14	5
4	29	19	10	97	70	27
5	43	32	11	37	24	13
6	32	26	6	26	20	6
7	112	88	24	45	32	13
8	32	22	10	53	44	9
9	34	28	6	64	50	14
10	32	25	7	62	44	18
	437	336	101	478	355	123

[The total (N) columns and the totals beneath the columns have been inserted by me.]

Later in his report, Mendel refers to his laws of segregation and independent assortment as *statistical* laws that hold on the average and are justified by large numbers of observations, but that may not hold when individual plants are considered. He was obviously impressed by the variation he observed among individual plants that, under his model, should have yielded 3 : 1 ratios.

Was Mendel's concern justified? The chi-square test can be used in arriving at a judgment. I won't perform the calculations here but merely state that a chi-square with 9 degrees of freedom can be calculated for each experiment in four easy steps (let the number of angular or green seeds be designated x):

1. Calculate the sum of x^2/N, summing over all 10 plants in each experiment (the answer is 24.18069240 for experiment 1).

2. Calculate the quotient of the (sum of x)^2 divided by the sum of N (23.34324943 in experiment 1).

3. Subtract item 2 from item 1 to get a difference ($d = 0.83744297$ in experiment 1).

4. Divide the difference by the product of 0.75×0.25, or 0.1875. The answer (4.47 in experiment 1) equals chi-square with 9 degrees of freedom.

The probability of seeing deviations as large or larger than those observed if each plant was in fact generating a 3 : 1 ratio of seeds equals 0.85 in experiment 1 and 0.65 in experiment 2. Thus, there is no indication in these data that individual plants varied in the ratios they produced. Like many people even today, Mendel seriously underestimated the meaningless variation that can occur by chance among samples of limited sizes. Variation among individual hybrid (i.e., heterozygous) plants beyond that expected by chance would have constituted a serious problem for Mendel's model—indeed, it may have defied explanation.

4 The Chromosome Connection

In his introductory remarks, Mendel referred to those who had preceded him: "Numerous careful observers, such as Kölreuter, Gärtner, Herbert, Lecoq, Wichura, and others, have devoted a part of their lives to this problem [the development of hybrids] with tireless persistence. Gärtner, especially, . . . has recorded very estimable observations." These were the plant hybridizers. Nowhere in his paper did Mendel refer to any microscopist. Despite Camerarius's plea to microscopists in 1694 that they help resolve the role of pollen in fertilizing eggs, Mendel made no comparable plea. He carried out his scrupulously careful pollinations without reference to any physical body that might possibly harbor the factors ("*Merkmalen*") responsible for the transmission of the characters he studied in his garden peas. Only a single remark reveals that Mendel did not regard heredity as a purely abstract matter: "The distinguishing traits of two plants can, after all, be caused only by differences in the composition and grouping of the elements existing in dynamic interaction in their primordial cells."

Perhaps "elements" was as far as Mendel could push the matter in 1865. J. A. Moore (1963) pointed out that the first description of the complex nuclear changes that occur during cell division was written (and illustrated) by A. Schneider (1873) almost accidentally during his research on flatworm morphology. Walther Flemming (1882), aided by advances in the chemistry of organic dyes and in the technology of optical instruments, made tremendous contributions to the understanding of chromosomal behavior during cell division. Boveri and Hertwig were instrumental in resolving chromosomal events, especially the halving of the normal (diploid) number of chromosomes with the result that each germ cell receives only the reduced (haploid) number.

Perhaps the search for the gene is best resumed by considering sex determination. In a study of a true bug (as opposed to *bug* used in the sense of "any *insect*"), *Pyrrhocris,* H. Henking noted that males possess 23 chromosomes. Twenty-two of these have proved to represent 11 pairs during spermatogenesis; the 23rd is unpaired. Furthermore, its behavior with respect to staining and movement during cell divisions is unlike that of the others. As an outcome of the reduction division that leads to sperm formation, two sorts of sperm are produced: those with 11 chromosomes and those with the same 11 *plus* the odd member (the X chromosome).

Nearly ten years later, after other such examples had been noted, Clarence E. McClung suggested that the X chromosome or its absence has something to do with sex determination. There exist two kinds of sperm in equal numbers and there exist two kinds of individuals: male and female; perhaps the two phenomena are related. Within the next five years, McClung's suggestion had been verified by E. B. Wilson and his students.

One of Wilson's students, W. S. Sutton (1903), after suggesting the previous year that the "association of paternal and maternal chromosomes in pairs and their subsequent separation during the reducing division . . . may constitute the physical basis of the Mendelian law of heredity," published a thorough analysis under the title "The Chromosomes in Heredity." His analysis proceeded as follows:

1. Each gamete carries a reduced number of chromosomes. Following fertilization, the cells of an individual carry pairs of chromosomes that constitute two corresponding chromosome sets.

2. During gamete formation, homologous chromosomes pair.

3. Following the cell divisions preceding gamete formation, each germ cell receives one member of each chromosome pair.

4. Different pairs of chromosome are separated independently of one another as far as paternal and maternal origins of the two members are concerned.

Sutton then went on to show that if Mendel's factors were parts of chromosomes, the reduction division would convert *Aa* cells into *A* and *a* gametes, with the two types being equally frequent. Furthermore, if *A* and *a* were parts of one homologous pair and *B* and *b* were parts of a second pair, *A B*, *A b*, *a B*, and *a b* gametes would be found in equal numbers. The first statement corresponds to Mendel's "law" of segregation, the second to his "law" of independent assortment (see Chapter 3).

Times arise during the development of a science when the solution to a problem virtually begs to be discovered. The rediscovery and appreciation of Mendel's findings (of his "laws," some would say) prompted many to seek a correspondingly orderly behavior of physical structures. Sutton was not alone in recognizing the parallels between chromosome behavior and the segregation and random assortment of maternal and paternal chromosomes; Theodor Boveri was another.

Boveri's 1902 study of dispermic eggs proved that the individual chromosomes are not equivalent to one another and that each has its own role to play in the normal development of an individual. Boveri added a footnote that A. H. Sturtevant (1965, p. 35) translated as follows: "I shall consider in another place these and related problems, such as the connection with the results of the botanists on the behavior of hybrids and their offspring." In his review of the history of genetics, Sturtevant (1965) also mentioned Karl Correns, who in 1902 deduced that segregation occurs during gametogenesis and even produced beads-on-a-string pictures of chromosomes with the beads labeled A, B, C, D, etc., in contrast with the alternative forms a, b, c, d, etc. About the same time, William Cannon and Michael Guyer published nearly simultaneous accounts in which Mendelian genes were said to be associated with chromosomes. "Thus," Sturtevant (1965, p. 37) said, "there were several people who were close to the correct interpretation at this time, but the first clear and detailed formulation was that of Sutton." And, later (p. 38), "At last, cytology and genetics were brought into intimate relation, and results in each field began to have strong effects on the other." Sturtevant did not use this term, but in 1902–1903 the new field of "cytogenetics" was born; it was the first of many new fields of biology in which "genetics" was attached to the name of an older, preexisting branch of biology. Genetics was eventually to provide the key to an understanding of the history of life, the development of individuals, and the moment-by-moment control of vital processes.

5 Clinching the Chromosome Theory: The Morgan School

Thomas Hunt Morgan, the embryologist who castigated early geneticists for bragging about the precision with which their data fit their theories, when in fact the theories had been developed to account for the very same data, became a geneticist himself during the early 1900s. He attracted to his laboratory (an extremely small room at Columbia University that was known affectionately as the "fly room" even while I was there during the 1940s) an extraordinary group of students and young colleagues: Calvin Bridges, H. J. Muller, and Alfred H. Sturtevant. These four, together with visiting geneticists from Europe, Asia, and elsewhere, sat virtually shoulder to shoulder at tables and desks placed against the walls of the room. Here they worked, speculated, discussed, and argued day after day; here, too, there was a joyous celebration each time a speculation was confirmed by experimental data or cytological observation. And here a bundle of dirty laundry lost by Calvin Bridges years earlier was found between his desk and the wall; Bridges died some years before this discovery.

The chromosome theory as expounded by Sutton and Boveri was accepted by nearly all geneticists, particularly by the Morgan group. Nevertheless, anomalies remained to be accounted for. The number of genes needed to explain the known Mendelian inheritance of simple characters, for example, greatly exceeded the number of chromosomes possessed by many organisms. For many years, to cite one example, geneticists referred to Mendel's luck in choosing seven characters for study—seven characters that exhibited independent assortment in a plant that possesses only seven chromosomes! The probability of choosing such characters by chance can be calculated as $\frac{7}{7} \times \frac{6}{7} \times \frac{5}{7} \times \frac{4}{7} \times \frac{3}{7} \times \frac{2}{7} \times \frac{1}{7}$, or 5,040/823,543 ($= 0.006$). This calculation, like those that emphasized the extraordinarily close fit of Mendel's observations to his theoretical expectations, prompted many to accuse him of doctoring his data. In fact, two of Mendel's genes lie on the same chromosome but are physically so far removed from one another that they exhibit independent assortment, a matter I discuss in this chapter. In fact, however, the probability of finding seven characters situated on six of seven chromosomes by chance exceeds 4%.

The test cross Mendel utilized to reveal the proportions of gene combinations carried by either pollen or egg cells involved matings with homozygous recessive plants (e.g., *aa bb cc dd . . .*). Because these plants form germ cells (*a b c d . . .*) containing recessive genes only, dominant genes contributed by the other parent are manifest in the offspring. A male parent whose genotype is *Aa Bb* should form germ cells of four sorts in equal proportions (*A B, A b, a B*, and *a b*); each of these types, when uniting with an egg of type *ab*, gives rise to an individual whose appearance differs from all others.*

A study of sweet peas by William Bateson and R. C. Punnett revealed the first reported instance in which F_2 progeny of a dihybrid cross did not fit the Mendelian expectation ($9:3:3:1$) even though each of the two characters taken individually gave a typical $3:1$ ratio. An example illustrating nonconforming F_2 data is shown here for round (*O*) versus elongate (*o*) fruit in tomatoes (as with Mendel's peas, the uppercase letter indicates dominance) and simple (*S*) versus compound (*s*) inflorescence (arrangement of flowers). The parental plants were true breeding (i.e., homozygous) elongate fruited with a simple inflorescence (*oo SS*), and round fruited with a compound inflorescence (*OO ss*). The F_1 hybrids bore round fruits in simple clusters (*Oo Ss*), a fact that reveals the dominance relationships. The F_2 generation can be represented as follows:

Phenotype	Number observed	Number expected if ratio were $9:3:3:1$
Round, simple	126	145.7
Long, simple	66	48.6
Round, compound	63	48.6
Long, compound	4	16.2
	259	259.1

*An important matter of terminology surfaces here. It is time to introduce (or reintroduce, because high school biology texts use this word) the term *allele*. The symbols *A, B, C,* and *D*

One scarcely need perform a chi-square calculation to see that these deviations are much greater than those expected by chance. (If this sentence is not obvious, the reader might want to actually carry out the calculations. With 3 degrees of freedom, a chi-square value of 7.82 would occur by chance in fewer than 1 instance out of 20, and a value of 11.34 fewer than 1 instance in 100. Note that the long, compound class alone contributes 9.19 [i.e., $12.2 \times 12.2 \div 16.2 = 9.19$] to the total chi-square, which, of necessity, must be larger than this single contribution.)

The discordant F_2 ratios illustrated above can be illustrated much better by the results of a test cross of the same F_1 hybrids to a doubly recessive strain (*oo ss*); in this case the four classes of offspring reveal the proportions with which gametes are formed by the hybrid parent:

Phenotype	Hybrid gamete	Number	Expected 1 : 1 : 1 : 1 ratio
Round, simple	*O S*	23	52.5
Long, simple	*o S*	83	52.5
Round, compound	*O s*	85	52.5
Round, simple	*o s*	19	52.5
		210	210.0

Clearly, the hybrid plant has not produced gametes in equal proportions; the parental combinations, *o S* and *O s*, greatly outnumber the newly arisen nonparental ones. The combinations of *O* and *o* with *S* and *s* have not arisen by chance. Furthermore, when the original parents were homozygous *OO SS* and *oo ss*, the F_1 hybrid (*Oo Ss*, seemingly the same as before) produced gametes of which 40% were *O S*, 40% *o s*, 10% *O s*, and 10% *o S;* the parental combinations once again exceeded the newly arisen combinations approximately four to one.

In these distorted ratios Bateson saw evidence that Mendelian factors segregated not at gametogenesis, as most people (including Mendel) had concluded, but at an earlier time, thus allowing for the reduplication (and multiplication) of certain combinations. Why the parental combinations should be favored (whether *O s* and *o S* or *O S* and *o s*, depending upon the parental cross) was not made clear; nor was a numerical relationship, (2^n −

refer to genes; we now know that genes lie on or are parts of chromosomes. The symbols *A* and *a* or *B* and *b* refer to contrasting forms (*allellomorphs,* or simply *alleles*) of genes. Thus, a chromosome consists of many genes that are located at as many different gene *loci* (places); at each *locus*, the gene may be represented by any one of its many *allelic* forms.

1) : 1, that Bateson postulated to fit the observed data ever explained. As is often the case in science, the explanation was offered by way of terminology—*repulsion* and *coupling*. These terms, however, merely provide names for what was observed; they explain nothing.

At this point, a return to chromosomal diagrams may be instructive. Although earlier I emphasized the reduction division—the division following which germ cells come to have only half (haploid) the number of chromosomes possessed by somatic, or body, cells (diploid)—to account for the constancy of both chromosome number and chromosomal material in successive generations of living organisms, I omitted an important

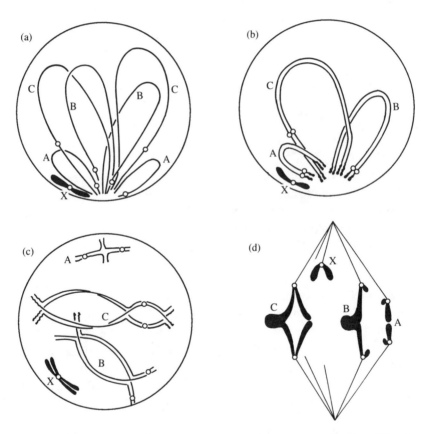

Figure 5-1. Meiosis in a hypothetical organism with three autosomes (A, B, and C) and an unpaired X chromosome. The exchange patterns shown in sketch c are discussed at length in the text. (Redrawn from White, 1948, by permission of Cambridge University Press.)

detail: the *chiasmata* (singular, *chiasma*) that hold the paired chromosomes together (see Figure 5-1).

Figure 5-1 represents the reduction division of an imaginary XO organism that possesses three pairs (A, B, and C) of autosomes. In diagram a, autosomes have reappeared as threads from an earlier, seemingly homogeneous, nucleus; the one X chromosome is precociously condensed. Each autosome "seeks" out its partner and, as if possessing a zipper, becomes closely attached to it (b); recall that each pair consists of a paternal and a maternal member. Just before metaphase the paired chromosomes take on the configuration shown in diagram c: Each chromosome, maternal and paternal, has become double structured. Furthermore, the two now-doubled chromosomes are held together by X-shaped regions where they seem to exchange partners. Without these points of exchange (*chiasmata*), the two chromosomes may fail to orient properly on the spindle and will fail to separate, one doubled chromosome to each pole.

What is the precise nature of the apparent exchange at the chiasmata? There are two possibilities. One is that the maternal and paternal chromosomes have been separated in different planes. Consider chromosome B in Figure 5-1c. If maternal strands are depicted as thin lines and paternal ones as much heavier ones, the diagram might be interpreted as follows:

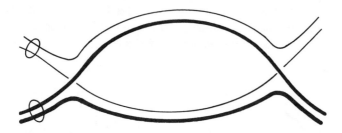

In this case, the separation of the two centromeres (the small regions, represented by circles, that are responsible for the movement of chromosomes during cell division) will yield:

That is, the maternal and paternal chromosomes remain intact.

Alternatively, chromosome B in diagram c might be represented as follows:

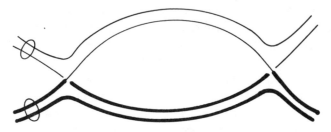

In this case, the separation of centromeres leads to the following:

Here, we see that a physical exchange of segments has occurred between maternal and paternal chromosomes.

Cytologists long debated which of the two interpretations was correct; I was offered both patterns as possibilities as late as 1941. Two special cases, however, provide nearly absolute proof favoring the exchange theory. The first concerns instances in which paternal and maternal chromosomes, although homologous, differ in size. Consider the case of a single exchange:

Notice that a separation in two planes with no exchange requires unmatched terminal pairings on the right side of the chiasma; these are never seen. The matched—long with long and short with short—configuration that is actually seen requires an exchange of paternal and maternal chromosomal material.

The second case occurs rarely because of errors committed during the "zipping" together of homologous (maternal and paternal) chromosomes before chiasma formation; interlocking bivalents (chromosome pairs) occasionally occur:

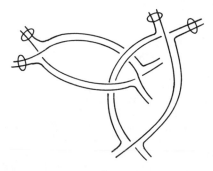

The nonexchange pattern illustrated here requires that one chromosome be able to pass through another without harm to either. No such difficulty occurs if the chiasmata are points of exchange between the two chromosomes of paternal and maternal origin.

Morgan, using the exchange model of chiasma formation as a basis, suggested that linkage merely reflects the proximity (or lack of it) of genes on the same chromosome. The low probability that an exchange would occur between genes whose locations were extremely close would lead to few recombinations between them; in contrast, genes at considerable distances from one another would, because of the high probability that an exchange would occur in the intervening region, show little or no linkage. We have already mentioned in another context that two of Mendel's "independent" factors are actually located on the same chromosome, but far from one another. (At any site along the chromosome, chromosomal material is exchanged between only two of the four strands. I leave it to the geometrically inclined reader to prove that, if the two strands involved in one chiasma are chosen randomly with respect to those involved in another, the amount of recombination between two genes, *A* and *B*, will never exceed 50%, precisely the value shown by genes undergoing independent assortment [see Figure 5-2]. This point was first demonstrated by a former teacher and friend, Marcus Rhoades of Indiana University.)

A. H. Sturtevant in his book *A History of Genetics* (1965, p. 47) adds this anecdote to linkage studies:

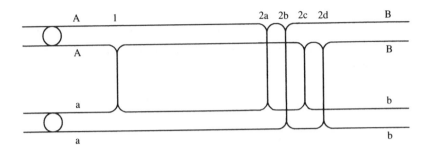

Figure 5-2. Chromosomal exchanges at two locations, 1 and 2, between the gene loci A and B. If only two of the four strands (chromatids) are involved at each location, and if the two strands involved at location 2 (2a, 2b, 2c, or 2d) are independent of those involved at location 1, then one can show by simple tabulation that the resulting crossover and noncrossover strands are equally numerous: 8 of each of a total of 16 strands. (For example, the 4 strands that result from crossovers 1 and 2a carry the alleles *A b, A B, a B,* and *a b*.) The complete tabulation is left to the reader. Note that during gametogenesis many instances occur in which *no* exchange occurs between the loci A and B; consequently, the proportion of noncrossover gametes among all gametes will always exceed that of crossovers.

In the latter part of 1911, in conversation with Morgan about [an earlier publication of William Castle of Harvard]—I suddenly realized that the variations in strength of linkage, already attributed by Morgan to differences in the spatial separation of the genes, offered the possibility of determining sequences in the linear dimension of a chromosome. I went home and spent most of the night (to the neglect of my undergraduate homework) in producing the first chromosome map, which included the sex-linked genes *y* [yellow body], *w* [white eye color], *v* [vermilion eye color], *m* [miniature wing], and *r* [rudimentary wings], in the order and approximately the relative spacing that they still appear on the standard maps.

Sturtevant is reputed to have said on a subsequent occasion that having made such a fundamental contribution to genetics as an undergraduate posed a severe problem for him later: How could he make a second, equally important contribution as an adult?

Nondisjunction: Aberrant inheritance patterns

The first mutant trait discovered in Morgan's laboratory after he and his students began studying the vinegar fly, or fruit fly, *Drosophila melanogas-*

ter, was white eyes. Rumor has it that Calvin Bridges spotted the small white-eyed fly, a spontaneous mutant, among hundreds of normal red-eyed flies in a bottle that he was about to place in a sterilizing oven. Whatever the circumstance of its discovery, the white-eyed male was saved from destruction and was used to start a genetic analysis.*

Historians often record only the successes in science, thus making it appear that our predecessors were intellectual giants who, unlike modern practitioners, never erred in their speculations. Thus, some readers may take pleasure in knowing that Morgan's first interpretation of the inheritance of white eyes was wrong; Moore (1963, pp. 80ff.) gives an analysis of the erroneous interpretation in which he describes the suppositions on which it was based and the consequences that would follow if it were true. Here, I describe only Morgan's second hypothesis; the one that proved to be correct.

When the *white* (-eyed) male was mated with a red-eyed (normal) female, all offspring, both male and female, had red eyes. When these F_1 red-eyed males and females were crossed, the daughters' eyes were red, whereas sons were divided: about half were red and half were *white.*

At the time the F_1 flies hatched, the original *white* male was still alive; consequently, he was mated with one of his red-eyed daughters. The outcome was an F_2 generation in which red- and white-eyed flies of both sexes appeared, and in approximately equal numbers.

Now, sex determination in *Drosophila melanogaster* is based on an XY mechanism (unlike the XO system in many grasshoppers that influenced McClung; see Figure 5-1): Male flies are XY, thus producing two types of sperm (X and Y) in equal proportions. Females are XX; they produce only X-bearing eggs. Fertilization of an egg by an X-bearing sperm results in a female (XX) individual; fertilization by a Y-bearing sperm results in a male (XY). The Y chromosome has remarkably few genes; the phenotype of male flies with respect to genes on the X chromosome is determined almost entirely by the single X chromosome they carry. Females, on the contrary, can be both homozygous and heterozygous for X-chromosome genes.

---

*T. H. Morgan (1942), correcting an erroneous account of the genesis of the white-eyed mutant, says, "The single white-eyed male that I found . . ." In this case, the "I" may signify "my laboratory" as an overreaction to an interloper's claim. The creative use of ambiguity (a common practice of scientists as well as other writers) is illustrated by Morgan's closing sentence: "Lutz closes his story with a characteristic and generous statement, 'If I had realized how valuable that white-eyed mutant was destined to be, I would not have been happy to give it away. However, it fell into good hands.' "

The inheritance pattern described above can be understood if the white-eyed male carried a recessive mutant gene (w) on its X chromosome (a *sex-linked* mutation). When mated with a normal female (WW, where W is the dominant allele that results in red eyes), daughters would be Ww (red eyed) and sons would be W Y (also red eyed). When the original white-eyed male (w Y) was mated with one of his daughters (Ww), both female and male offspring would be of two types: Ww (red) and ww (*white*) daughters, and W Y (red) and w Y (*white*) sons. This interpretation was confirmed by numerous subsequent crosses; indeed, the white-eyed mutant is a favorite in the genetics classroom because even beginning students can easily recognize red and white eyes.

A mating of red-eyed males (W Y) with white-eyed females (ww) should produce red-eyed daughters (Ww) and white-eyed sons (w Y). As a rule, that is the outcome. Nevertheless, if many hundreds of such offspring are raised, an occasional white-eyed female and, equally seldom, an occasional red-eyed son appear. The red-eyed sons in this case are sterile; the white-eyed females are fertile. If the latter are crossed with normal red-eyed males, exceptional white-eyed females and red-eyed sons are found in greater numbers (about 4% of all progeny are exceptional); the red-eyed males in this case are fertile.

Calvin Bridges interpreted these events correctly. The exceptional white-eyed females in the first instance require two X chromosomes (in order to be female), both of which must carry the recessive allele w. The male cannot contribute to these females (because the X-bearing sperm carries the dominant, normal W allele). Thus, during oogenesis, an occasional egg must be produced that contains two X chromosomes, an error called *nondisjunction*. Equally frequent are eggs with no X chromosomes. After fertilization, the following possibilities exist:

Sperm	Exceptional eggs	
	X^wX^w	O
X^W	$X^WX^wX^w$ Dies	X^WO Red-eyed male
Y	X^wX^wY White-eyed female	YO Dies

The exceptional male flies without a Y chromosome are sterile and red eyed; the exceptional white-eyed female flies carry a Y chromosome which (unlike the case for human beings and other mammals) affects neither their fertility nor their sexual phenotype.

When these exceptional white-eyed females (*ww* Y) are mated to normal males (*W* Y), the frequency of nondisjunction (secondary as opposed to the primary nondisjunction discussed above) is increased; the cause is the mispairing of an X with the Y chromosome rather than the usual X-with-X pairing that gives rise to normal disjunction. Finally, in the case of secondary nondisjunction, the exceptional red-eyed males are fertile because they now possess a Y chromosome that they have obtained from their XXY mothers.

Bridges not only worked out this hypothetical scheme to account for the exceptional males and females but also prepared squashes of their ovaries and testes for cytological examination. In every instance, his expectations were borne out: males obtained following primary nondisjunction lacked Y chromosomes; exceptional females proved to be XXY—that is, they carried a Y chromosome in addition to their two Xs.

Bridges (1916) published his results in a paper titled "Non-disjunction as Proof of the Chromosome Theory of Heredity." Today, it is accepted as proof and referred to as such in nearly all genetics textbooks. The reason is clear: an exceptional phenotypic inheritance pattern proved to be accompanied by an equally exceptional chromosomal inheritance pattern. To cite an analogy, one accuses two students of cheating if they submit identical *incorrect* answers, not if both have arrived at the correct one. Proof resides in the identity of the *unexpected*. Nevertheless, Morgan (1940) had this to say in an obituary of Calvin Bridges, who died at the relatively young age of 49:

A much more extended paper appeared in 1916 entitled "Non-disjunction as Proof of the Chromosome Theory of Heredity." This work, offered as his doctoral dissertation at Columbia University, included not only genetic evidence but corresponding evidence from a study of the chromosomes that tallied with the genetic results. This paper went far towards convincing skeptics and conservatives that chromosomes are the bearers of genetic factors. It is true there was abundant evidence before 1916 showing that chromosome behavior furnishes an interpretation of heredity. It is today hard to believe that *it was nearly ten years before this relation was generally accepted*. (Emphasis added)

Giant chromosomes

Scientists are a thrifty lot, especially when it comes to time. They utilize (within the bounds of ethics) whatever means will provide answers to their questions most rapidly and with the least effort. An understanding of sex linkage made the X chromosome of *Drosophila melanogaster* a favorite tool among early workers; the discovery of nondisjunction (followed by the discovery of a female fly whose two X chromosomes were attached to the same centromere, and hence of necessity were transmitted from a mother to her daughters as a single unit) provided splendid opportunities for searching out new mutations and even determining the rates at which such mutations occur.

While Morgan and his group were working on *Drosophila melanogaster* at Columbia University and were proselytizing geneticists at other universities to do the same, a husband-and-wife team, Phineas and Anna Whiting, were studying the genetics of a small parasitic wasp (*Habrobracon*). Like all hymenopterans—wasps, bees, hornets, and ants—males are haploid, whereas females are diploid. Unfertilized eggs develop into males (drones, in bee colonies), while fertilized eggs develop into females. (There is a persistent rumor that Mendel developed his ideas knowing that a hybrid [German × Italian] queen bee produces male progeny one-half of whom are German and one-half Italian. A fellow beekeeper [Mendel was an amateur beekeeper] and clergyman had published results stating that the two types of males appeared in a 1 : 1 ratio [Whiting, 1935].) The haplodiploid sex-determining mechanism in *Habrobracon* means that all chromosomes in this organism behave like sex chromosomes; that is, they behave like the X chromosome in *Drosophila*. The ease with which mutations could be obtained in *Habrobracon* merely by screening males made this organism extremely attractive for certain types of genetic studies. It actually threatened to displace *Drosophila melanogaster* as the preferred research organism in many laboratories.

The vinegar fly (fruit fly) was saved as an organism for genetic research by a most timely discovery. Although they were observed as early as 1881, large nuclear structures that occur in certain cells of Diptera were recognized as giant chromosomes only in 1933. T. S. Painter, a cytogeneticist (and later university president) at the University of Texas made the first study of these structures in the larval salivary glands of *Drosophila melanogaster*. The visible banding patterns were constant enough from larva to larva—and sufficiently different from chromosome to chromosome—to allow him to make detailed maps of the entire chromosome set.

Calvin Bridges, with his extraordinarily keen vision—Sturtevant acknowledges that Bridges had the best eye for new mutants as well—soon took over the construction of giant chromosome maps. His cytological procedures and cataloging system are still standard for those who prepare comparable maps for poorly studied *Drosophila* species or for other dipteran species. Figure 5-3 illustrates the X chromosome of *Drosophila melanogaster* when represented as a giant chromosome dislodged from the nucleus of a salivary gland cell (Bridges's map) and (at the same magnification) as a member of an oogonial metaphase plate. The large size of the salivary chromosome results, apparently, from its failure to condense in preparation for cell division (salivary gland cells no longer undergo division) and from 10 or 11 duplications and reduplications that produce 1024–2048 parallel chromosomal strands, all tightly apposed to one another. The basal third of the mitotic X chromosome is not represented in the salivary chromosome; it fails to undergo the repeated divisions and, consequently, is represented in the salivary gland cell nucleus only as a small tangled mass at the base of the large (euchromatic) chromosome. A close look at the drawing of the giant X chromosome shows that it consists of two closely united structures; these structures are the two chromosomes, each composed of 1000–2000 replicated strands. The larvae from which preparations were made were females (XX). Somatic pairing of homologous chromosomes is characteristic of nearly all dipteran flies.

Figure 5-4 represents an early use to which salivary chromosomes were put. Imagine a wild-type, red-eyed male that has been exposed to X-radiation; since 1927 it has been known that such radiation can cause gene mutations, chromosomal rearrangements, and losses (or, less frequently, duplications) of chromosomal material. This irradiated male is mated to white-eyed (*ww*) females. Among the many red-eyed (*Ww*) daughters that result there will also be a few white-eyed ones. These white-eyed females are mated to males that carry a normal X chromosome, and resultant female larvae that carry the irradiated X chromosomes are examined cytologically. The argument in this experiment is that one obtains a white-eyed female daughter of an X-rayed red-eyed male only if the normal gene for red eyes (*W*) has been mutated or physically lost. The black bars in Figure 5-4 represent independently obtained losses of X-chromosomal material (referred to as deletions or deficiencies) that remove the *W* locus (thus reveal-

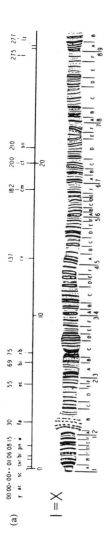

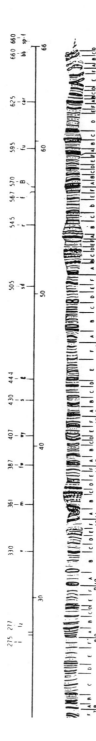

(b)

Figure 5-3. The contrast in size between a giant chromosome of the sort found in the salivary gland nuclei of *Drosophila melanogaster* (a) and the same chromosome (drawn to the same scale) as seen in a female oogonial cell (b). Only the giant X chromosome is depicted; it extends from its tip (section 1) to its base (section 20); section 8 has been shown in both upper and lower portions of the figure. The X chromosomes of the oogonial (egg-forming) cell are the two rod-shaped ones at 11:00–12:00 o'clock in the small figure. (From Bridges, 1935, by permission of *Journal of Heredity*.)

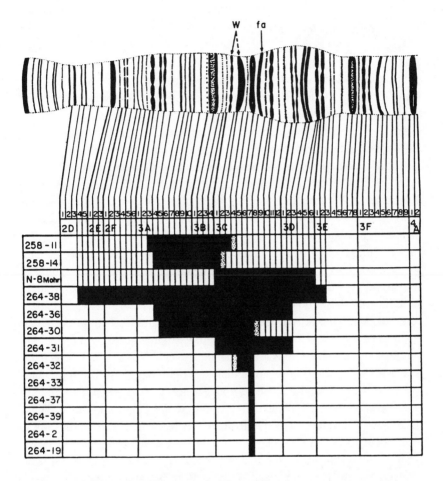

Figure 5-4. The use of overlapping deficiencies to determine the physical location of genes relative to individual bands of giant salivary gland chromosomes.

The chromosomal segment shown here is part of the X chromosome of the vinegar fly, or fruit fly, *Drosophila melanogaster*. The numbers in the left-hand column designate those chromosomes, obtained experimentally, from which certain bands are missing; the black bar to the right of each number indicates the missing bands.

Deletions 258-11 through 264-31 are known (from genetic tests) to remove the gene for white eyes. Consequently, this gene is on, or adjacent to, band 3C1, the only band common to these 7 deficiencies. The 11 deficient chromosomes, from N-8 through 264-19, remove a second gene known as *facet* (mutations at this locus disrupt the structure of the eye). This gene must be on or near band 3C7, a band missing from all 11 of these deficient chromosomes.

Occasionally a chromosome seems to have a deletion on the basis of a genetic test, but microscopic examination fails to reveal it. This need not disturb us unduly because (1) human eyes are not infallible, (2) microscopes, even the best ones, have optical limitations, and (3) invisible gene mutations caused by alterations at the molecular level may produce effects indistinguishable from those of deficiencies. (Modified from Wallace, 1966.)

ing the mother's recessive *w* allele); hence, the band or bands lacking from each radiation-induced white deletion must include the one containing the white locus. The figure illustrates that relatively few deletions are required to pinpoint the locus of a particular gene to one (or at most a few) band(s). Figure 5-5 illustrates how chromosomal deficiencies and duplications actually appear under microscopic examination of giant salivary chromosomes.

A final point can be made here that bears on the conclusions Wilhelm Roux arrived at in 1883. Small chromosomal deletions such as those illustrated in Figures 5-4 and 5-5 have been recovered throughout the genome of *Drosophila melanogaster*. Many of these are quite large—including as many as 70 bands, for example. Many that are listed in the compendium of *Drosophila* mutations (Lindsley and Grell, 1968) include 50 or more bands. These deletions are recoverable and analyzable, of course, because their heterozygous carriers survive and reproduce. Matters are quite otherwise, however, with respect to individuals that are homozygous for deletions. All the deletions we have mentioned so far are lethal when homozygous. Homozygous lethality extends downward to include deficiencies only 3–5 bands long. Indeed, several deletions too small to be detected by the microscope are lethal when homozygous. The only deletions that the compendium lists as viable when homozygous are extremely small ones (2–3 bands) located at the extreme tips of chromosomes. Thus, of the 5162 bands represented in Bridges's maps of the giant chromosomes, nearly every one is essential for the survival of a developing embryo, larva, pupa, or adult fly. The notion that each band represents one gene has been an attractive one for geneticists. Wilhelm Roux's surmise about the impor-

---

Figure 5-5. The pairing of giant chromosomes that differ in the presence or absence or duplication of small segments. (a) The pairing of giant chromosomes, one of which lacks several bands (arrow). Outside the region of this small deletion, pairing is perfect; at the site of the deficient segment, the bands of the normal chromosome (those within the indicated arc) form a loop because there is nothing with which they can pair. It is important to note that the bands without pairing partners do not pair with one another; pairing attraction is only between corresponding bands, *not* between otherwise unpaired bands. (b) The pairing configuration of two chromosomes, one of which carries a duplicated segment. The accompanying line drawing clarifies the pairing pattern. The duplicated bands simply seek out their counterparts and pair with them; this gives rise to a short segment in which there are three matching chromosome parts rather than the usual two.

These examples offer proof that bands related in origin—and only bands related in origin—pair in the precise manner observed in giant chromosomes. (Modified from Wallace, 1966.)

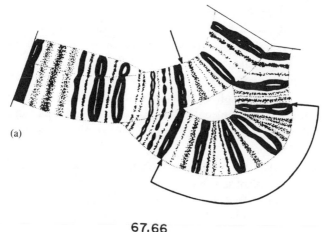

(a)

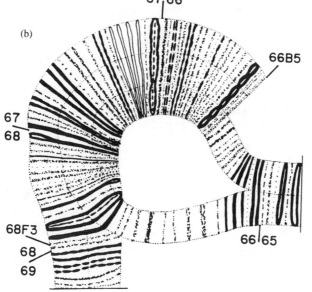

(b)

67 66

66B5

67
68

68F3

68

69

66 65

tance of dividing each particle of chromosomal material during each cell division so that both daughter cells receive their portion is absolutely correct: there is little or no functional redundancy among microscopically visible particles of genetic material of most higher animals. These carefully chosen words are necessary because molecular biologists have discovered much repetition and duplication at their level of analysis. It is as if, in studying a watch, the molecular scientists discovered a plethora of cogs; despite the seeming sameness of cogs, every small gear is needed if the watch is to function.

In summary, we have now traced the gene to its location on the chromosome; indeed, the giant salivary chromosomes of *Drosophila melanogaster* have allowed geneticists to locate hundreds of genes (two dozen are indicated in Figure 5-4) either within individual bands (of which there are more than 5000 in Bridges's maps) or in small regions that include only several bands. Geneticists, incidentally, are accomplishing much the same with human chromosomes; more and more genes are being located on not only one or the other of the 23 chromosomes but also within one of the dozen or more identifiable regions within individual chromosomes (Figure 5-6).

Let me make a final point in concluding this chapter. Boveri showed by means of his analyses of dispermic eggs that each chromosome has its own function during normal development. Roux emphasized that each chromatin particle must be divided and shared by daughter cells. These observations imply that there is no "healing" at the chromosomal level, that a missing chromosome cannot be replaced by material contributed by another. This suggestion is verified in a most spectacular way by the aberrant chromosomes listed in the compendium of *Drosophila* mutations. Chromosomes lacking small pieces such as those illustrated in Figures 5-4 and 5-5 do not heal; the missing pieces have been missing in some instances for 75–80 years. The lesions first noted in these chromosomes are the ones still present today. In 1855, Rudolf Virchow claimed that all cells arise from preexisting cells; by 1882, cytologists were claiming that all nuclei come from preexisting nuclei and that all chromosomes come from preexisting chromosomes. What we have learned of persistent deficiencies in salivary chromosomes allows us to join Roux and claim that each chromosomal band arises from an earlier band—generation after generation. Furthermore, no band can be either physically or functionally replaced by even a close neighbor.

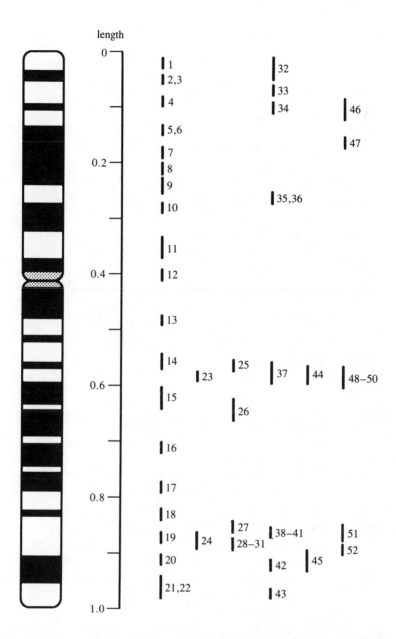

Figure 5-6. Chromosome 11 of the human genome. Although not nearly as detailed as a dipteran's giant salivary gland chromosome, 29 clearly definable regions are visible. To the right are indicated the approximate positions of chromosomal regions that hybridize with one or another of 52 different DNA probes. (From Fearen et al., 1990, copyright 1990 by the AAAS.)

Box 5-1. Scientific models and their supporting data

H. G. Andrewartha, one of Australia's outstanding ecologists, in explaining his science to beginning students, commented on models and model making as follows: "Model-making, the imaginative and logical steps which precede the experiment, may be judged the most valuable part of scientific method because skill and insight in these matters are rare. Without them we do not know what experiment to do. *But it is the experiment which provides the raw material for scientific theory* [emphasis added]. Scientific theory cannot be built directly from the conclusions of conceptual models" (1963, p. 181).

Contrast that statement with Sturtevant's frank admission: "One of the striking things about the early Drosophila results is that the ratios obtained were, by the standards of the time, very poor. With other material it was expected that deviations from the theoretical Mendelian ratios would be small, but with Drosophila such ratios as $3:1$ or $1:1$ were rarely closely approximated. This was recognized as being due to considerable differences in the relative mortalities of the various classes in larval and pupal stages, before counts were made" (1965, p. 51). The poor correspondence between actual data obtained by the Morgan group and the models they were said to support was not lost on Bateson (1916, p. 540). "For instance, we find the following extraordinary series given," he said, and then he listed seven sets of numbers from which map distances had been estimated by the Morgan group—numbers among which it is impossible to recognize the expected patterns of two equally numerous recombinant classes and two equally numerous parental ones. "The machinery [possessed by the Morgan group] for dealing with unconformable cases is extraordinarily complete," he continued (p. 541). This machinery included differential viability, association of alleles with lethal factors, changes with age, a special phenomenon called "interference," and "an altogether distinct kind of crossing-over in the four-stranded stages."

Another person inclined to place greater importance on experimental observations than on theoretical models was W. E. Castle of Harvard University. He was skeptical about the linear linkage map that the Morgan group (especially Sturtevant) was constructing. He doubted, for example, "whether an elaborate organic molecule ever has a simple string-like form." Furthermore, distance AC should equal distance AB plus distance BC (where A, B, and C represent three gene loci); however, AC is almost always less than the sum of AB plus BC, especially if distances AB and BC are substantial. Thus, B cannot lie on a line connecting A and C but, rather, must lie to one side. Last, by adding crossover distances, Morgan's group obtained total distances exceeding 50%. This, Castle claimed, was impossible because random assortment equals 50% recombination; linkage, therefore, must always result in fewer than 50% recombinants. As a result of his reinterpretation of Morgan's data, Castle produced the cat's-cradle-like figures (the "rat-trap" model in Morgan's terminology) shown in Figure 5-7b.

Some of Castle's errors can be anticipated; one has already been exposed. Both DNA and polypeptide chains (the basis of protein molecules) are, in fact, linear molecules of great length; these points are emphasized in a subsequent chapter. In my account of chiasmata and the recombination of genes, I emphasized that, at any one site, only two of the four strands are involved (this was Bateson's "distinct kind of crossing-over"); furthermore, I implied (leaving the proof largely to the reader) that as long as recombination involves only two of the four strands at each site, recombinants cannot exceed 50% no matter at how many sites recombination occurs between two gene loci. In fact, map distance as computed by the Morgan group equals one-half the average number of sites at which crossing-over occurs between any two genes.

In response to Andrewartha's comments, I must point out that the chromosome theory rests on a simple model; ecological theory does not. Slight disturbances in temperature, rainfall, or the introduction or loss of some form of life (a predator, herbivore, or pathogenic organism) can have enormous and totally unexpected consequences in natural communities. Such uncertainties underlie most of the present-day controversy surrounding global warming, environmental deterioration, the fate of endangered species, and the release of genetically engineered organisms into the environment.

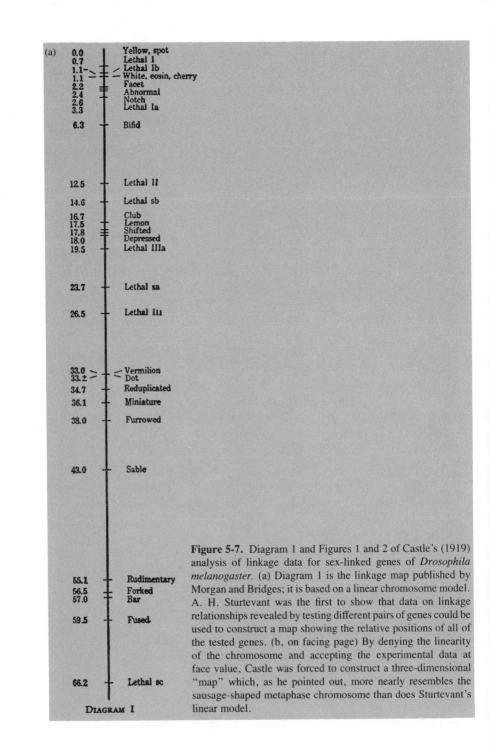

(a)

0.0	Yellow, spot
0.7	Lethal 1
1.1	Lethal lb
1.1	White, eosin, cherry
2.2	Facet
2.4	Abnormal
2.6	Notch
3.3	Lethal Ia
6.3	Bifid
12.5	Lethal II
14.6	Lethal sb
16.7	Club
17.5	Lemon
17.8	Shifted
18.0	Depressed
19.5	Lethal IIIa
23.7	Lethal sa
26.5	Lethal IIi
33.0	Vermilion
33.±	Dot
34.7	Reduplicated
36.1	Miniature
38.0	Furrowed
43.0	Sable
55.1	Rudimentary
56.5	Forked
57.0	Bar
59.5	Fused
66.2	Lethal sc

DIAGRAM I

Figure 5-7. Diagram 1 and Figures 1 and 2 of Castle's (1919) analysis of linkage data for sex-linked genes of *Drosophila melanogaster.* (a) Diagram 1 is the linkage map published by Morgan and Bridges; it is based on a linear chromosome model. A. H. Sturtevant was the first to show that data on linkage relationships revealed by testing different pairs of genes could be used to construct a map showing the relative positions of all of the tested genes. (b, on facing page) By denying the linearity of the chromosome and accepting the experimental data at face value, Castle was forced to construct a three-dimensional "map" which, as he pointed out, more nearly resembles the sausage-shaped metaphase chromosome than does Sturtevant's linear model.

(b)

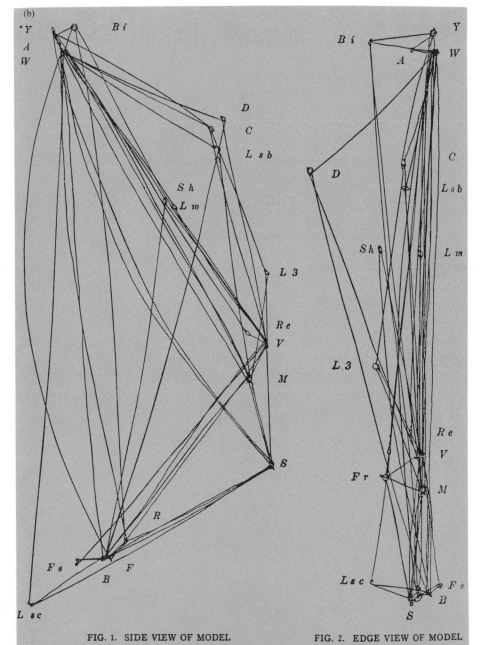

FIG. 1. SIDE VIEW OF MODEL

FIG. 2. EDGE VIEW OF MODEL

Seen at right angles to figure 1.

Box 5-2. Cytological proof of genetic crossing-over

If genes lie on or are parts of chromosomes, and if the movement of chromosomes during gametogenesis accounts for the Mendelian segregation and assortment of genes, then it should follow that recombination of genes is accompanied as well by the recombination of chromosomal material. So reasoned four geneticists about 1930: Curt Stern of the University of California, Harriet Creighton and Barbara McClintock of Cornell University, and R. A. Brink of the University of Wisconsin. The studies these four carried out are so direct and simple (in retrospect) that their inclusion in the present text would require at most two paragraphs. The anecdotes told in reference to these three studies provide a view of science that is generally unavailable to an outsider; hence this boxed account.

To reveal that actual material is exchanged between two chromosomes when genes recombine requires that the chromosomal material be visibly marked; recall that one of the difficulties encountered by those searching for the gene was that all stained mitotic chromosomes look alike. Indeed, Bateson (1916, p. 542) wrote: "The supposition that particles of chromatin, indistinguishable from each other and indeed almost homogeneous under any known test, can by their material nature confer all the properties of life surpasses the range of even the most convinced materialism." By now, we know that Bateson was wrong, but in 1930 proof was still desirable.

Chromosomes can be physically labeled in a number of ways. In corn, chromosomes carry "knobs" (globs of chromatin) of various but consistent (for a particular chromosome) shapes and sizes. Translocations—physical exchanges between chromosomes—were known for both corn and *Drosophila melanogaster* in 1930. Hence, one could prepare in either experimental organism true-breeding stocks of the following sorts:

$$\begin{array}{ccc}
\blacktriangle \dfrac{A \quad B}{\quad} \bullet & & \dfrac{a \quad b}{\quad} \\[2mm]
& \text{and} & \\[1mm]
\blacktriangle \dfrac{A \quad B}{\quad} \bullet & & \dfrac{a \quad b}{\quad}
\end{array}$$

On the left, two different physical markers flank the genes A and B; on the right, no such markers are present in the true-breeding $a\ b$ strain.

The F$_1$ hybrid obtained by crossing these stocks can be represented as

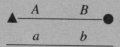

When F$_1$ hybrids are crossed with the doubly recessive strain, four kinds of test-cross progeny are obtained whose genetic and chromosomal constitutions are shown below.

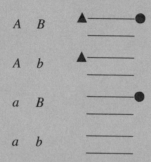

The observed patterns of chromosomal markers are precisely what all three studies revealed: chromosomal segments are indeed exchanged when nearby genes form new combinations.

Creighton and McClintock published their results on corn in 1931, the same year Stern published his more extensive and elaborate study on *Drosophila melanogaster*. Fifty years later, many ex-Cornellian geneticists gathered at Cornell for the 75th anniversary of the Synapsis Club, a social club initiated originally by Cornell's plant breeders. At this meeting, among many other reminiscences, was the following account of the Creighton-McClintock study related by Creighton.

T. H. Morgan came to Cornell to visit his friend and colleague R. A. Emerson, an outstanding corn geneticist and, at that moment, a professor with an exceptional group of students and postdoctoral fellows (Figure 5-8). As is customary on such occasions, Morgan spent time discussing research problems with the graduate students. When he learned of Creighton and McClintock's results, he insisted that they be published. The preliminary results, however, were exceptionally mea-

Figure 5-8. Professor R. A. Emerson (with cap) and his student colleagues at Cornell University in 1929. Standing with Emerson, from the left, are C. R. Burnham, M. M. Rhoades, and Nobel laureate (1983) Barbara McClintock. Posing with the Emerson dog is the late George Beadle, also a Nobel laureate (1958). (Photograph courtesy of the Department of Plant Breeding, Cornell University.)

ger because the growing season for corn in Ithaca had been terrible that summer. Morgan was not fazed: "Publish your results," he insisted.

"But next year's crop will be larger; suppose the additional results do not agree with what we have now?" The two students were quite concerned.

"Why, then you must publish those results," Morgan replied. "In that case, you will have had two publications where you should have had none."

Curt Stern was not amused by the early publication of Creighton and McClintock's results. In fact, upon meeting the pair at a subsequent scientific meeting, he was noticeably cool toward them. Once more Morgan entered the scene. Sensing the hostility, he stepped between Stern and the two women. "Curt," he said to the former, "you be nice to these two ladies. They work with corn and get only one generation per year. You and I work with flies and get a generation every two weeks." His argument seems to have carried the day.

With R. A. Brink matters were quite different. Because he was intent upon gathering sufficient data to make his case foolproof, his experiments on corn were still in progress when Creighton and McClintock published their results. When he published several years later (1935), Brink pointed out with some feeling that the preliminary data published by Creighton and McClintock were inadequate to clinch the point they were attempting to make. (Note that ambiguous results would have been obtained if recombination occurred between ▲ and *A*, or *B* and ●, in the earlier diagrams.) Creighton and McClintock's subsequent reply (1935) reveals the low level of statistical sophistication characteristic of biologists of that era. "Brink has accused us of having insufficient data to make our case in our earlier publication," they said in effect. "We would now like to present additional data to show that we did have enough earlier." They may have clinched their case with their additional data, but this has no bearing on the question concerning the adequacy of the preliminary data—an adequacy about which they expressed doubts themselves when first meeting with Morgan.

As a footnote to this episode, I find that drosophilists generally refer to Stern when discussing the exchange of chromosomal material during recombination; botanists and botanical geneticists tend to favor Creighton and McClintock. The importance of publishing early is emphasized by a total lack of references to Brink; only by reading Creighton and McClintock's second paper did I learn of his contributions to this problem.

6 Chromosome Chemistry

At the conclusion of Chapter 5 we found the gene firmly located in the chromosome. The giant chromosomes of *Drosophila melanogaster* even revealed the positions of genes with respect to individual bands. Figure 5-4 shows the locations of two genes—*white* (eye color) and *facet* (eye structure)—in two bands separated by only three others, the entire complex being only 1/1000th of the total number of visible bands in such chromosomal preparations.

As small as most chromosomes may be, they are enormous in molecular terms. Where within these sausagelike structures do genes actually reside? Of what are they made? Is a chromosome merely a pockmarked body with a gene residing in each small crater when it isn't out regulating metabolism in the cytoplasm? That is one of the more fanciful suggestions made by geneticists before 1920.

To appreciate the difficulties besetting the search, one must appreciate the complexity of organic chemistry. The term *organic* itself reveals a great deal: organic chemicals were so named because, although they were routinely made by living organisms, their synthesis lay far beyond the skill of laboratory chemists. At least, that was the view until Friedrick Wöhler synthesized urea in his laboratory in 1828. Subsequent progress, although steady, has been (by modern standards) painfully slow.

Proteins

Proteins are, of all organic molecules, the most complex and the most likely to be referred to as "living." A standard histology textbook of the

1930s, *Bailey's Text-book of Histology,* the one I used as an undergraduate, says outright, "The living protoplasm of the cell is a slimy semifluid or viscid substance." Contrast that statement with one written by James D. Watson: "Cells contain no atoms unique to the living state; they can synthesize no molecules which the chemist, with inspired hard work, cannot some day make. There is no special chemistry of living cells" (1965, p. 68). Proteins are the main constituent (except for water) of cellular slime. These large molecules are composed of building blocks called amino acids. One of the first amino acids to be discovered was cysteine, in approximately 1810. Ninety years later, in 1900, 16 amino acids were known. There are 20 altogether, not counting modified versions (see Figure 6-1; Table 6-1 lists the names of the amino acids and their abbreviations). The complexity of their structure is responsible for the lapse of a century before scientists understood how they are built. The role of each in a given protein molecule is a matter of research even in the 1990s.

The chemical linkages that unite amino acids into the chain that is the protein molecule (Figure 6-2) were known by 1900, the year Mendel's paper was rediscovered. By the early 1940s, specific proteins were known to consist of constant proportions of amino acids. The molecular weights of particular proteins (such as trypsin) were often estimated by expressing the proportions of amino acids in a manner such that each could be represented by an integer (one, two, three, . . .); no amino acid could be present in the protein molecule as a fraction (e.g., one and one-half molecules of lysine). As far as I can recall, the linear array of amino acids itself was not also regarded as being constant for molecules of a given protein. Indeed, in 1938, Dorothy Wrinch argued strongly (if not convincingly for all protein chemists) that the linear structure postulated for proteins was a degradational artifact; that cross-linkages other than disulfide bonds (—S—S—) provided globular structures, which in turn formed protein fabrics; and that synthetic polypeptide chains made by linking amino acids together in the laboratory lacked the properties of real protein molecules.

The search for the gene that we have undertaken (and to which many people preceding us dedicated their lives) need not be converted into a treatise on organic chemistry. It must suffice to state that the postulated single-file, linear arrangement of amino acids (polypeptide chain) proved to be the correct model for protein structure—the *primary* structure, that is. Added to this is the molecule's *secondary* structure—a helical one (not the grapes-on-a-vine construction shown in Figure 6-2) in which stability is generated by chemical attractions between molecules exposed at turns of

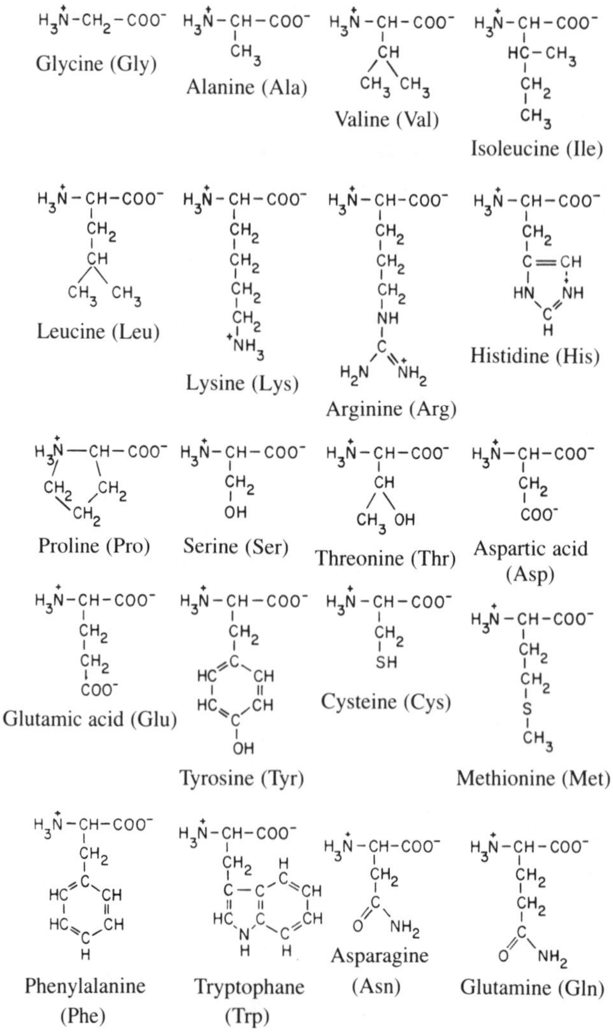

Figure 6-1. The 20 amino acids (names, chemical structure, and conventional symbol) essential for the synthesis of proteins of living organisms. An outstanding discovery of modern biology (perhaps *the* outstanding discovery) was the revelation that proteins such as those found in meat or eggs have precise structures. Each protein consists of a unique and accurate sequence of these amino acids—as many as 10,000 or more amino acid building blocks in the larger protein molecules. Fortunately, this discovery coincided with the discovery of the structure of deoxyribonucleic acid (DNA), the chemical basis of heredity; otherwise, the means for achieving precision in the synthesis of molecules at this level of structural complexity would have been a mystery.

Table 6-1. The 20 amino acids involved in protein synthesis together with their three-letter and one-letter abbreviations (the latter are used in Figure 12-2).

Three-letter abbreviation	Amino acid	One-letter abbreviation
Ala	Alanine	A
Arg	Arginine	R
Asn	Asparagine	N
Asp	Aspartic acid	D
Cys	Cysteine	C
Gln	Glutamine	Q
Glu	Glutamic acid	E
Gly	Glycine	G
His	Histidine	H
Ile	Isoleucine	I
Leu	Leucine	L
Lys	Lysine	K
Met	Methionine	M
Phe	Phenylalanine	F
Pro	Proline	P
Ser	Serine	S
Thr	Threonine	T
Trp	Tryptophane	W
Tyr	Tyrosine	Y
Val	Valine	V

Figure 6-2. An old diagram showing several arginine molecules linked to form a polypeptide chain. The linkage between amino acids is chemically correct; each involves the removal of OH from —COOH and H from H₂N— (thus, the removal of a molecule of water) to form the chemical link. Physically, the diagram is incorrect. The peptide linkages tend to force the protein chain into a self-reinforcing helical structure; amino acids do not dangle from a polypeptide chain like clusters of tropical fruits. This primary structure is generally disrupted by other forces, however, that impinge on the protein molecule.

the helix. (Linus Pauling, two-time Nobel laureate, proposed the alpha helix as the basic structure of a polypeptide chain; higher-order interactions occurring within protein molecules distort this basic helical structure.) The *tertiary* structure of the protein molecule involves foldings and turnings that are stabilized by disulfide cross-linkages between rather widely spaced molecules of the sulfur-containing amino acid cysteine. Finally, protein molecules may consist of collections of two or more polypeptide chains rather than one.

A final, genetically oriented account of protein structure can be presented at this time. Using procedures developed by Frederick Sanger during the 1940s, several workers determined the amino acid sequences in several proteins by breaking the proteins into fragments. Sanger's technique can be illustrated by disintegrating the first sentence of this paragraph. Suppose I have an enzyme that cuts sentences only after the letter *n,* in whatever word it may occur. Following the exposure of the sentence to that enzyme, I would have the following fragments floating around:

A fin	be presen
ted accoun	al, gen
etically orien	t of protein
structure can	ted at this time

Now, suppose I have a second enzyme that cuts words only after *b* and *t;* it would generate these fragments from the original sentence.

A final, genet	of prot
ein st	ruct
ed account	ically orient
ure can b	e present
ed at	t
his t	ime

Letters within short fragments can be deciphered. I leave it to the reader to prove that the two sets of fragments can be used in reconstructing the original sentence. And so it was with protein molecules: by determining the sequence of amino acids in the small fragments of a particular protein—fragments, however, that were generated by two different proteolytic enzymes—workers could deduce the amino acid sequence of the entire protein (Figure 6-3). The important outcome of such research was the

Hydrolytic Agent **Amino Acid Sequence**

Hydrolytic Agent	Amino Acid Sequence
Trypsin	Ser.Tyr.Ser(Met.Glu.His.Phe)Arg
Chymotrypsin	Arg.Trp
Trypsin	Trp.Gly.Lys.Pro.Val.Gly.Lys
Trypsin	Lys.Arg
Trypsin	Lys.Arg.Arg
Acid	Arg.Pro.Val.Lys
Trypsin	Pro(Val,Lys,Val,Tyr)
Acid	Val.Tyr.Pro.Ala.Gly.Glu(Asp.Asp.Glu.Ala.Ser.Glu.Ala.Phe.Pro.Leu.Glu.Phe)
Pepsin	Ala(Gly.Glu.Asp)
Pepsin	Asp.Glu
Pepsin	Asp(Glu.Ala)
Pepsin	Asp(Glu.Ala)Ser
Pepsin	Glu(Ala.Ser)
Pepsin	Ser.Glu
Pepsin	Ser(Glu.Ala)
Pepsin	Ser(Glu.Ala.Phe)
Pepsin	Glu.Ala.Phe
Pepsin	Phe(Pro.Leu.Glu)
Pepsin	Pro(Leu.Glu.Phe)

Complete Sequence: Ser.Tyr.Ser.Met.Glu.His.Phe.Arg.Trp.Gly.Lys.Pro.Val.Gly.Lys.Lys.Arg.Arg.Pro.Val.Lys.Val.Tyr.Pro.Ala.Gly.Glu.Asp.Asp.Glu.Ala.Ser.Glu.
1 2 3 4 5 6 7 8 9 10 11 12 13 14 15 16 17 18 19 20 21 22 23 24 25 26 27 28 29 30 31 32 33

Ala.Phe.Pro.Leu.Glu.Phe.
34 35 36 37 38 39

Figure 6-3. The reconstruction of a polypeptide hormone (α-corticotropin) from the peptide fragments obtained by proteolytic digestion or acid hydrolysis. Note that sequences of ambiguous order (listed in parentheses) are clarified by nonambiguous sequences from other fragments. The complete sequence is given at the bottom of the figure. (From Anfinsen, 1959, by permission of Christian B. Anfinsen.)

knowledge that a given protein does indeed consist of a precise, constant sequence of amino acids.

During the first half of the twentieth century two red blood cell disorders that came to be known as sickle-cell anemia and sickle-cell trait were discovered; they were later studied by J. V. Neel as a familial characteristic. After some initial confusion that resulted from the usual (at that time) early death of anemic children, the disease was found to be the result of an incompletely recessive gene, *s*. *SS* and *Ss* individuals are "normal" under most circumstances; *ss* individuals suffer from severe anemia and a number of often fatal anemia-related ailments. Red blood cells of *Ss* individuals (who are said to exhibit the sickle-cell trait) become misshapen (hence the term *sickle*) if these cells are deprived of oxygen (as on a microscope slide coated with the dye methylene blue). During World War II it was learned that red blood cells in the spleens of *Ss* individuals flown at high altitudes in unpressurized airplanes also sickle; the spleens of these persons were likely to rupture during flight.

Linus Pauling and his colleagues demonstrated that the hemoglobin (Hb) of *Ss* individuals is of two types that separate if placed in an electric field

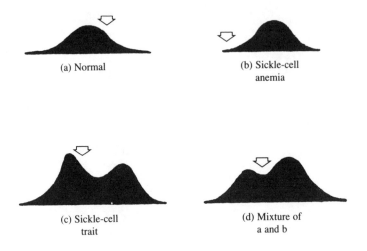

(a) Normal

(b) Sickle-cell anemia

(c) Sickle-cell trait

(d) Mixture of a and b

Figure 6-4. The migration of normal (a) and sickle-cell (b) hemoglobins in an electrical field. Note that the hemoglobin of heterozygotes with the sickle-cell trait (c) behaves much like a mixture of two types of hemoglobin (d): normal hemoglobin and that of homozygotes suffering from sickle-cell anemia. Arrows indicate the position of the original spot of hemoglobin (before migration) and can be used as points of reference. (Redrawn from Pauling et al., 1949, copyright 1949 by the AAAS.)

(see Box 6-1). The two types moved in different directions (toward different [+ or −] poles) under the conditions of their experiment. The hemoglobin of *SS* persons migrated toward the positive pole, that of *ss* individuals migrated toward the negative pole, and that of *Ss* persons showed two components, one of which migrated in one direction while the other migrated in the opposite direction (much like mixtures of hemoglobin from *SS* and *ss* persons; see Figure 6-4).

The difference in migration patterns of normal and sickling hemoglobins was later shown to arise from a single amino acid substitution. Using a combination of chromatographic and electrophoretic separations of protein fragments obtained by the digestion of normal and sickling hemoglobins, V. M. Ingram demonstrated that only 1 of 26 fragments differed in its relative positions on the two-dimensional patterns created by the two hemoglobins (Figure 6-5). This fragment (by chance, the leftmost end of the hemoglobin molecule) consisted of eight amino acids:

	Amino acid site							
	1	2	3	4	5	6	7	8
Normal Hb	Val	His	Leu	Thr	Pro	*Glu*	Glu	Lys
Sickle Hb	Val	His	Leu	Thr	Pro	*Val*	Glu	Lys

Thus, a single-gene abnormality of hemoglobin proved to consist of the substitution of one amino acid (valine) for another (glutamic acid). This change altered the electrical charge of the molecule, thus allowing Pauling and his colleagues to detect the changed molecule in an electric field (and also allowing them to coin the phrase *molecular disease*); because the change involved only one amino acid, it affected only 1 of the 26 fragments Ingram obtained by digesting hemoglobin with a proteolytic enzyme.

Nucleic acids

Nucleic acid was discovered in 1869. That is a historical fact, although no publication appeared for another two years. The discoverer was Friedrich Miescher; his experimental material was pus cells he had obtained from discarded hospital bandages (antisepsis and especially antibiotics were not part of the hospital environment at the time).

According to F. H. Portugal and J. S. Cohen (1977), Miescher set out to

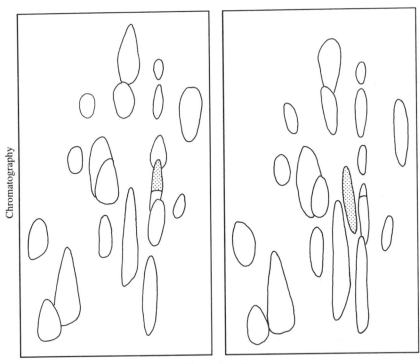

Chromatography

Electrophoresis

Figure 6-5. Diagrams ("fingerprints") of the positions to which fragments of hemoglobin move on a sheet of paper when exposed to chromatography in one direction and electrophoresis in another. On the left is the pattern created following the digestion of normal hemoglobin; on the right, that of sickle-cell hemoglobin. Only one fragment has altered its position; analysis of the amino acids in that fragment reveals that only a single amino acid has been changed: the glutamic acid in position 6 of normal hemoglobin has been replaced by valine. This single change is also responsible for the different mobilities illustrated in Figure 6-4. (Redrawn from Ingram, 1963, by permission of V. M. Ingram.)

study proteins but quickly encountered another substance that could not belong to any protein class known at the time. The new substance proved to be what is now known as DNA (deoxyribonucleic acid).

Once more I will resist the temptation to recite a litany of chemical advances. The important point here is that there proved to be two types of nucleic acids, not one. The nucleic acid obtained readily from calf thymus glands and fish sperm proved to be DNA. The nucleic acid from yeast cells proved to be a related but not identical substance, RNA (ribonucleic acid).

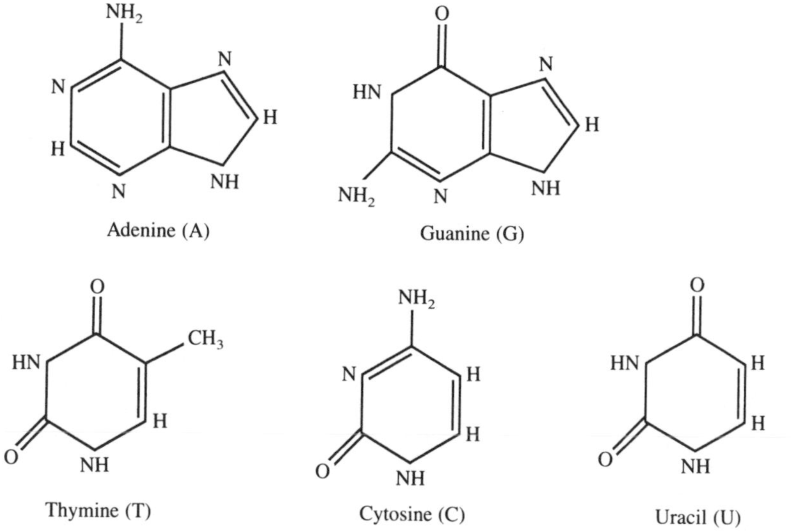

Figure 6-6. The chemical formulas for the five purine and pyrimidine bases found in nucleic acids. Adenine and guanine are purines; thymine, cytosine, and uracil are pyrimidines. The first four named are found in deoxyribonucleic acid (DNA); uracil replaces thymine in ribonucleic acid (RNA).

Further research proved that these large molecules (frequently referred to as *macromolecules,* together with proteins and complex carbohydrates) are constructed of subunits (see Figure 6-6): guanine (G), cytosine (C), adenine (A), thymine (T), and uracil (U). Only the first four (G, C, A, and T) are found in DNA; U replaces T in RNA. A five-carbon sugar (pentose) is an essential part of both nucleic acids, but again, the two differ (Figure 6-7): one atom of oxygen is lacking in the sugar molecule of DNA relative to that of RNA (hence, *deoxy*ribonucleic acid). Finally, both nucleic acids contain ample quantities of phosphoric acid—hence that term in both names. As was true with respect to proteins, workers with different views debated whether these complex substances were truly chemical molecules with definite structures or merely colloidal aggregates whose proportions varied from particle to particle.

Following the discovery of a staining procedure (the Feulgen test) specific for DNA and a procedure for isolating cell nuclei, DNA was found to exist only in nuclei, confined to chromosomes. (The small circular bits of DNA now known to exist in the cytoplasmic organelles—mitochondria

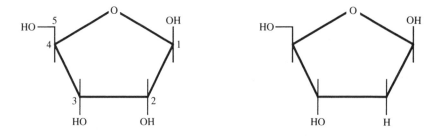

Figure 6-7. The two pentose (five-carbon) sugars of the nucleic acids. At the left is ribose, the sugar component of ribonucleic acid (RNA). At the right is deoxyribose (notice the lack of an oxygen atom at position 2), the sugar component of deoxyribonucleic acid (DNA). In Chapter 12 still another variant, dideoxyribose, is mentioned; it lacks the oxygen atom at position 5 as well. This variant is useful in the laboratory because, without the oxygen atom in position 5, the synthesis of a filamentous DNA molecule is unable to proceed; this has been exploited in DNA sequencing because the incorporation of a dideoxyribotide halts the elongation of a growing DNA molecule. Nucleotides containing dideoxyribose are thus known as terminator molecules.

and chloroplasts—are not visible under the light microscope, even when stained by the Feulgen procedure.) Furthermore, the amount of DNA per cell was found to be constant for a given species regardless of the tissue from which the cells were obtained, with one exception: sperm cells and egg nuclei contain only half the amount of DNA present in diploid body cells (see Table 6-2).

Despite the better understanding biochemists gained of the chemical composition of chromosomes, the physical nature of the gene remained a matter of controversy. The physical structure of DNA was largely unknown until early 1953. By 1950, most of the following facts were known about DNA (after Moore, 1963):*

1. It consists of six kinds of molecules: adenine, thymine, guanine, cytosine, deoxyribose, and phosphoric acid.

2. Physical measurements suggested that DNA molecules were huge (polymers or macromolecules).

*Items 3 and 4 of this list were not known in 1950. Aaron Klug, writing in the Norton Critical Edition of Watson's *The Double Helix* (Stent, 1980), states that by February 1952, Rosalind Franklin knew that there were very likely two chains in the DNA helix she was analyzing by X-ray diffraction. Dr. Franklin's analyses also led to the conclusion that the helix is 20 Å in diameter.

Table 6-2. The DNA content (grams $\times$ 10^{-12}) of various cell types from different animals. At the top are listed early data which emphasize that sperm cells contain only one-half the amount of DNA found in the animal's body cells, which are diploid. During spermatogenesis the diploid number of chromosomes is reduced by half, thereby also reducing the amount of DNA by half. The full amount of DNA is restored at fertilization when diploidy is also restored. At the bottom of the table are more recent data that emphasize the enormous range of DNA amounts per cell in different organisms.

Organism	Kidney	Liver	Erythrocytes	Sperm
Chicken	2.4	2.5	2.5	1.3
Cow	6.4	6.4	—	3.3
Carp	—	3.0	3.3	1.6
Human being	5.6	5.6	—	2.5
T2 bacteriophage			0.0004	
Escherichia coli			0.010	
Sponge (*Dysidea crawshagi*)			0.1	
Tunicate (*Ciona intestinalis*)			0.4	
Coelenterate (*Cassiopeia*)			0.7	
Amphioxus (*Amphioxus lanceolatus*)			1.2	
Echinoderm (*Lytechinus pictus*)			1.8	
Flatfish (*Pleuronichthys verticalis*)			1.6	
Trout (*Salmo gairdneri*)			5.6	
Mouse (*Mus musculus*)			7.0	
Frog (*Rana pipiens*)			15.2	
Newt (*Triturus viridescens*)			91.0	
Salamander (*Necturus maculosus*)			195.2	
Lungfish (*Lepidosiren paradoxa*)			247.8	

Note: — = not determined.

3. DNA seemed to consist of two fibers twisted about one another in the form of a helix, like the ascending and descending ramps in a parking garage.

4. The diameter of the helix is 20 Å.

5. Each fiber consists of alternating phosphate-deoxyribose-phosphate-deoxyribose components.

6. The cytosine, thymine, adenine, and guanine units (also known as purine and pyrimidine bases) are attached to the fibers.

7. In different cells of the same species, the relative amounts of the four bases are constant.

8. From species to species, the proportions of the bases can vary greatly.

9. In all cells, whatever their source, the proportions of guanine and cytosine are equal, as are the proportions of adenine and thymine. Because

adenine and guanine are purines, and thymine and cytosine are pyrimidines, these two types of molecules—purines and pyrimidines—occur in
equal amounts.

To these essentially modern "hard" facts, we might add the following
comments by Muller (1961, p. 7): "It is an illusion to think that without
having this unique replicational property any organization like protein or
protoplasm could gradually become more and more anabolic, in the sense
of producing a structure having exactly its own complexities. For there is
no way by which any variation in it, no matter how conducive to its growth

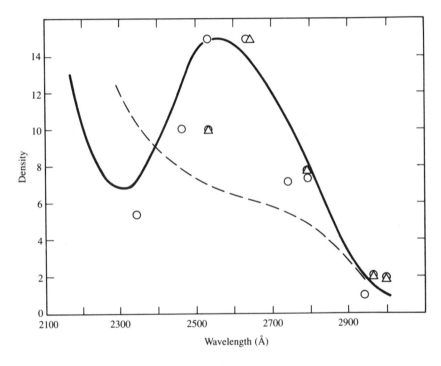

Figure 6-8. The absorption spectra of DNA and protein in the ultraviolet (UV) region of
electromagnetic radiation. Superimposed upon the spectra are the relative frequencies with
which gene mutations were induced in two experiments (Stadler and Uber, 1942; Knapp and
Schreiber, 1939) by UV radiation of different wavelengths. Note that the experimental results
follow the DNA absorption spectrum quite well while departing considerably from the
protein absorption curve. In retrospect, these observations seem to support rather strongly the
hypothesis that DNA, not protein, is the genetic substance. (O = Stadler and Uber, △ =
Knapp and Schreiber; solid line = DNA, dashed line = protein.)

in general, could come to produce more of its own distinctive features and thus to share in its growth and multiplication." Muller was summarizing views expressed in the early 1920s: hereditary material must be not only able to replicate; it must also, when altered, be able to replicate the altered form as precisely as the original. In this earlier era, Muller (1929) became virtually despondent: "Just how these genes thus determine the reaction-potentialities of the organism and so its resultant form and functioning, is another series of problems, at present practically a closed book in physiology, and one which physiologists as yet seem to have neither the means nor the desire to open" (quoted in Muller, 1962, p. 196).

Thus, we close this chapter having learned that, despite their complexities and intractable natures, much was known about proteins and deoxyribonucleic acid (two organic materials that compose the bulk of plant and animal chromosomes) by the late 1800s and early 1900s. Nevertheless, by the mid-1900s there was no compelling argument capable of convincing everyone concerned that either protein or DNA was *the* genetic material (see Figure 6-8).

Box 6-1. Chromotography and electrophoresis

The axes in Figure 6-5 illustrating the difference between normal and sickling hemoglobin in fragment 4 are labeled "chromotography" and "electrophoresis." What, at least in essence, do these terms mean? What techniques did Ingram use in this study?

Chromotography (paper chromotography) is not difficult to understand. I have just now proven that to myself. Before sitting down to write, I took a paper towel from the kitchen, cut it in half, and (using a set of colored pens that I keep for visiting children) drew a half dozen or more differently colored lines 1 inch long about 2 inches from and parallel to the bottom of the towel. Using Scotch tape, I hung the toweling from a cupboard so that its lower edge just touched the surface of water in an aluminum pie tin. As the water rose in the toweling, it passed through the variously colored lines, causing them to run and smearing them upward.

The paper toweling, now dry, is in front of me. The pigments have run at different speeds, and I can easily see that several of the colors I chose to test are really composites. Four of them contain the same fast-moving yellow component; three have a slow-moving purple component that lies near the original line; pink and blue pigments are also present in several of the colors I tested. Paper chromotography, of which I have just cited a poor man's version, is the separation of components of a complex mixture by virtue of their different diffusion rates in a solvent. The polypeptide fragments in Ingram's hemoglobin digest were separated vertically by the passage of a solvent (not necessarily water) through the paper on which a spot of material had been placed in the lower left-hand corner.

The left-to-right separation obtained after the chromatography was achieved by electrophoresis. A household demonstration of this technique is difficult because it requires direct—not alternating—electrical current. The procedure also works better at high voltage.

Of the 20 amino acids that make up proteins, 16 have uncharged "side" chains, 2 bear positive electrical charges, and 2 carry negative ones. These 4 amino acids provide the electrical charges that lead to the migration (just as ions migrate in an electrified salt solution) of proteins in an electrical field. In a protein containing 200 amino acids, approximately 10 of each of the 4 charged amino acids might be expected (some, of course, will be rarer; others more common); thus, there may be 20 positively charged ones and 20 negative ones. Proteins (or polypeptide fragments) in acidic buffer solutions tend to migrate toward the positive pole at a rate largely determined by the total electrical charge. Pauling and his colleagues chose a buffer that actually resulted in the movement of the two types of hemoglobin molecules in opposite directions.

Box 6-2. One gene–one enzyme

In this search for the gene I have forsworn extended accounts of gene action; gene action is a chemical phenomenon best left to those conversant with biochemical reactions. On the other hand, there is the ancient admonition, By their fruits shall ye know them. Knowledge of the gene may be forthcoming from an understanding of what genes do.

The relationship between gene and morphological (phenotypic) character was stressed by Mendel (smooth versus wrinkled seed, yellow versus green seed color, tall versus short plants, and still other segregating traits) and, later, by the *Drosophila* workers of the Morgan group. It was up to two others, George Beadle and Boris Ephrussi, to experiment further with mutant fruit flies. Using two mutants that (superficially, at least) have identically bright red eyes (*cinnabar* and *vermilion*) and wild-type flies whose eyes are dark red because of the presence of a brown pigment that both bright-red-eyed mutants lack, Beadle and Ephrussi transplanted larval tissues destined to become adult eyes from one larva to another. Tissues were transplanted from wild-type (+) larvae to larvae of all three sorts (+, *cn, v*), from *cn* larvae to all three sorts, and from *v* larvae. The host larvae grew, pupated, and finally emerged as adult flies. Lying free in the abdomen of each adult host was a third eye, one that arose from the transplanted tissue.

The results of the Beadle-Ephrussi experiment are shown in Figure 6-9. On the indicated diagonal (upper left to lower right) are transplants between larvae of the same sort; transplanted eyes resemble the host's eyes in every case. The top row reveals that the wild-type host provides something (some material substance, since we are not mystics) to the transplanted *vermilion* and *cinnabar* eyes that enables them to achieve normal coloring (i.e., to make the otherwise missing brown pigment). The left-most column reveals that cells destined to become wild-type eyes do so even if transplanted into larvae of mutant flies; indeed, the eyes of the mutant (*cn* and *v*) host adults tend in these cases to be somewhat darker than the usual mutant eye.

The exciting outcome is revealed by the normal eye color achieved by cells from *vermilion* larvae that were transplanted into *cinnabar* flies, and the complete absence of any such effect in the opposite case in which *cinnabar* eye-forming cells were transplanted into *vermilion* larvae. This outcome suggests that the formation of eye pigment (brown) can be ascribed to a series of chemical reactions:

$$A \rightarrow B \rightarrow \ C \rightarrow \text{brown pigment}$$
$$\uparrow \quad \uparrow$$
$$v^+ \quad cn^+$$

Cinnabar (*cn*) and *vermilion* (*v*) flies lack brown pigment because each is unable to carry out the reaction that is controlled by its wild-type (normal) allele (superscript +). *Cinnabar* and *vermilion* eye-forming cells develop into normal eyes in wild-type larvae because these larvae

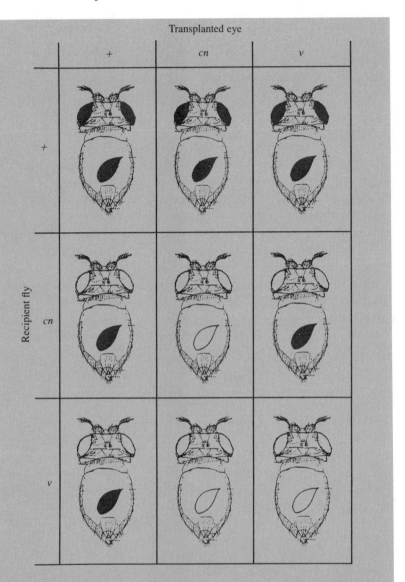

Figure 6-9. The outcomes obtained by transplanting preadult eye tissue (eye *anlagen*) from either wild-type, *cinnabar*, or *vermilion* larvae into "host" larvae (also wild-type, *cinnabar*, or *vermilion*), which were then allowed to grow, pupate, and emerge as adult *Drosophila melanogaster*. These adults, in addition to their own two eyes, contain a third adult eye within their abdominal cavities. Transplants from *vermilion* larvae into *cinnabar* hosts result in wild-type (dark) third eyes; those from *cinnabar* larvae to *vermilion* hosts do not. The explanation for this critical observation is given in the text.

provide both substances, B and C; that is, these are diffusible substances (hormones were suggested by Beadle and Ephrussi). *Vermilion* eye-forming cells make dark-red eyes in *cinnabar* larvae because although *v* cells cannot make B they can convert B into C if B is provided by the host. On the other hand, *cn* eye-forming cells remain bright red in *vermilion* larvae because these larvae can provide neither B nor C to the *cinnabar* eye-forming cells.

Drosophila flies are complex higher organisms. Beadle and Ephrussi could only assume that the materials labeled v^+ and cn^+ were enzymes necessary for the synthesis of the diffusible substances B and C. Although the phrase "one gene–one enzyme" (meaning that each enzyme—a protein—requires a normal gene for its synthesis) was used following the *Drosophila* transplantation experiments, it came into prominence after this type of biochemical research was transferred to *Neurospora,* the pink bread mold. Here is an organism that requires only some inorganic salts, a carbon source such as sugar for energy, and a ubiquitous vitamin (biotin) for growth and reproduction. Every complex organic substance it needs (the "living slime" referred to earlier) is made by the mold itself—a remarkable ability that makes *Neurospora* extremely useful in genetic research.

Beadle, in collaboration with E. L. Tatum, irradiated *Neurospora* spores (see Figure 6-10) and subsequently germinated sexual spores individually in tubes containing a complete medium (one containing amino acids, vitamins, and other organic substances); the mold colonies were then tested, individually in separate tubes, on minimal medium. *Neurospora* that failed to grow on the latter were screened for growth on media enriched, say, by a collection of vitamins or a mixture of amino acids. Next, having seen which mixture permitted growth, the colony was tested once more in a series of tubes, each containing minimal medium enriched by a single amino acid or a single vitamin (Figure 6-10, bottom). In the vast majority of cases in which a mutant was unable to live on minimal media, that inability was overcome by adding a particular chemical (vitamin or amino acid) or a closely related one. In each of these cases, subsequent genetic tests showed that a single gene was involved in the inability to synthesize the chemical.

Mutant genes of this sort have proven extremely useful to biochemists as well as to geneticists. A collection of mutants that cannot grow in the absence of D, for instance, may prove upon finer analysis to be of

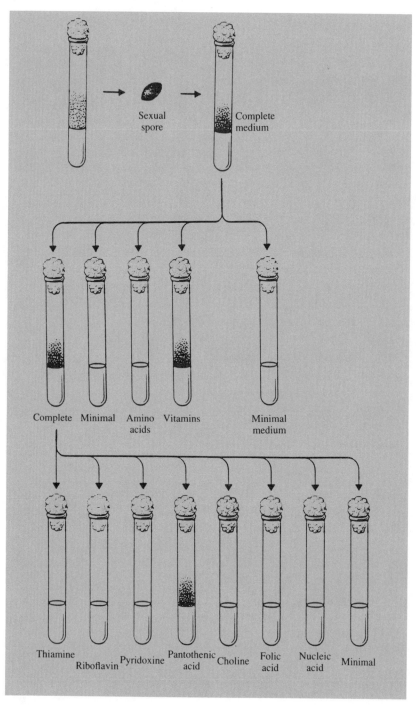

several sorts: those that require B, others that require C, and the rest, which grow only in the presence of D itself. From such information one can infer that the underlying metabolic pathway can be represented as:

$$A \rightarrow B \rightarrow C \rightarrow D$$

$$\begin{array}{ccc} \uparrow & \uparrow & \uparrow \\ \text{1st} & \text{2d} & \text{3d} \\ \text{group} & \text{group} & \text{group} \end{array}$$

It was such information in the case of the colon bacterium, *Escherichia coli,* that led not only to a biochemical pathway such as that shown here but also to the genetic information (obtained by gene recombination) that the genes responsible for specifying the needed enzymes are themselves organized *in the same physical sequence*. This sort of assembly-line arrangement of enzymes is not commonly found in organisms higher than bacteria.

Box 6-3. The origin of bacterial genetics

When asked recently to define *geneticist,* a colleague replied, "One who studies inheritance by making crosses"; all other practitioners are cell biologists, biochemists, or molecular biologists. Had he been reminded, he probably would have identified Mendel as the first geneticist in order to exclude a multitude of earlier plant hybridizers who recorded general observations concerning the hybrids they produced but no quantitative data on the transmission of specific, easily scored, alternative traits.

Figure 6-10. The experimental procedure used by Beadle and Tatum to isolate radiation-induced mutations in *Neurospora* and to demonstrate that each of these "biochemical" mutants is unable to carry out one—and only one—metabolic transformation. From these studies emerged the "one gene–one enzyme" hypothesis: each normal gene is responsible for the synthesis of one, and only one, enzyme. The tube on the upper left contains *Neurospora* obtained by mixing wild-type conidia of two mating types, one of which has been exposed to mutagenic radiation. The sexual spore (and the mold that grows from it) is haploid; that is, it carries only one representative (allele) at each gene locus. Although the mold in this illustration can grow on a complete nutrient medium, it cannot grow on one (minimal medium, center right) consisting only of a few organic salts and a carbon source. A further test (center left) reveals that its growth requires the addition of vitamins to the growth medium. Finally (bottom), pantothenic acid is shown to be the specific vitamin required for growth. Stated differently, this mutant strain of *Neurospora* is unable to synthesize pantothenic acid from simple precursors.

For much of its existence the science of genetics was restricted to studying the transmission of heritable differences through successive generations. One could proclaim with impunity that even without heritable variation certain branches of biology would exist: If all cats were genetically and phenotypically identical, there would still be a cat *anatomy,* a cat *physiology,* a cat *embryology,* and a cat *neurobiology.* But there would be no cat *genetics.* J. B. S. Haldane alluded to this contradiction when he said that he could easily explain the difference between a black cat and a white cat but that he would have great difficulty in explaining their overall similarities.

When geneticists turned their attention to the bacterium *Escherichia coli* as a possible experimental organism, they were unable to perform crosses as they had done previously in corn, *Drosophila,* and even *Neurospora.* Bacteria with altered characteristics could be isolated as individual mutants. The frequency with which mutants arose could be calculated and compared with frequencies of gene mutations in other organisms. Novel combinations of unusual phenotypes could be obtained by mutation alone. For example, *a b* individuals could be obtained from *A B* individuals by two routes:

$$A B \rightarrow A b \rightarrow a b$$
$$A B \rightarrow a B \rightarrow a b$$

To confirm, however, that these changes are genetic in the sense that heritable differences in higher organisms are genetic was impossible.

Joshua Lederberg, who recently retired as president of Rockefeller University, was largely responsible for converting the studies of *E. coli* from purely phenotypic to truly genetic ones. Lederberg realized that sexual crosses between pairs of bacteria may occur at extremely low frequencies and hence might go undetected in most studies. A procedure was needed by which the progeny of these (hypothetical) rare matings would be revealed; that procedure of necessity must either destroy or immobilize all nonsexually reproducing bacteria.

Lederberg's procedure relies on the ability of wild-type *E. coli* to synthesize all its constituent chemical compounds from some inorganic salts, a simple nitrogen compound, and an organic carbon source for energy. Some mutants are unable to synthesize this or that of the hundreds (even thousands) of chemicals needed for bacterial growth and reproduction; these mutant individuals must be provided with these particular chemicals—amino acids, vitamins, or other vital compounds. The mutants, that is, grow on a chemically defined minimal medium only if it has been enriched with a specific compound (see Figure 6-10 for a comparable situation in *Neurospora*).

Lederberg's analytical procedure was deceptively simple. A mixture of bacteria, half of which required substances a and b for growth, while half required substances c and d, were spread on petri dishes containing only a chemically defined minimal medium with no enriching substances. If uppercase letters designate the ability of an organism to synthesize a given substance, the mixture of bacteria created by Lederberg can be represented as:

$$a\,b\,C\,D \times A\,B\,c\,d$$

None of these bacteria is able to grow (i.e., to form colonies) on minimal medium; only $A\,B\,C\,D$ organisms can grow without chemical supplements. But Lederberg did recover such colonies! Thus, *E. coli* became amenable to standard genetic analyses: frequencies of recombination could be computed and genetic maps could be constructed. As is so often the case, the study performed by Lederberg (in collaboration with E. L. Tatum; both received the Nobel Prize for their work on recombination in *E. coli*) succeeded because one of the strains (K-12) used was capable of transferring its chromosome to the other strain. Not all mixtures of bacterial strains would have yielded recombinant progeny capable of growing on minimal medium.

7 Molecules of Authority

In previous chapters we have seen Mendel's Merkmal (factor) evolve into Bateson's and Morgan's gene. Further, evidence has accumulated that the gene is housed in the chromosome. First, Mendelian inheritance patterns and those of chromosomes can be superimposed (Sutton and Boveri), even when the latter are abnormal (Bridges, nondisjunction of sex chromosomes). Recombination of genes is accompanied by corresponding recombination of nearby physical markers on chromosomes (Creighton and McClintock, Stern). Deletions of small portions of chromosomes of *Drosophila* have proved to be visible in the giant salivary chromosomes, thus allowing genes to be located on individual bands.

Despite all of the above, we are still left with the question, What chemical component of the chromosome represents the genetic material? Proteins are complex substances; life is complex. DNA seems too simple. But how about the action spectrum of ultraviolet-induced mutations illustrated in Figure 6-8? To have an effect, radiation must be absorbed; sunlight heats black plastic, it does not heat transparent Saran Wrap. The figure clearly shows that the effect of ultraviolet light (in terms of gene changes or gene mutations) closely follows the DNA absorption curve, not that of protein.

No, the evidence is not enough to remove attention from the more complex, the more "living" protein (remember the 1931 reference to "living protoplasm" [p. 87]). Muller, for example, suggested that by absorbing UV radiation, DNA transfers the absorbed energy to the proteinaceous gene. That suggestion was not entirely nonsensical. About the

108

time it was made, Hugo Fricke, a biophysicist residing on Long Island, New York, showed that the inactivation of enzyme molecules by X-radiation was not caused by direct "hits" on the molecules; if it were, the proportion of molecules inactivated would be independent of the concentration of the aqueous enzyme solution exposed to radiation. Enzyme inactivation results from molecules of water (H_2O) that have been split into uncharged H (atomic hydrogen) and OH (hydroxyl) radicals. These extremely reactive radicals transfer the energy of radiation to the enzyme molecule, thus bringing about further chemical changes that leave the enzyme incapable of reacting with its substrate.

Obviously, the search for the gene had to continue.

Naked molecules

Pneumococcus (now known as *Streptococcus pneumoniae*) is the bacterium responsible for pneumonia; at one time this disease was the leading cause of death in the United States. Infection with pneumococcus is especially severe in mice; inoculation with a single pneumococcus bacterium inevitably results in a fatal infection. There are, however, two forms of this bacterium, *rough* (R) and *smooth* (S), so called because of the appearances of colonies grown on culture medium in petri dishes. The difference in appearance is caused by a polysaccharide (complex sugar) coat that surrounds each individual bacterium. The rough individuals lack the polysaccharide coating and are incapable of causing an infection in either mice or humans.

The polysaccharide coat eventually proved to be of several constructions (originally designated types I–IV; today many more types are known), which could be recognized using the sera of persons who had or had had pneumonia. Infected persons develop antibodies against surface molecules of infecting organisms; such antibodies provide the body's natural resistance to disease. Thus, the antibodies formed in response to an infection by pneumococcus type S-I are anti-S-I; those formed in response to type S-II are anti-S-II. Anti-S-I sera cause bacteria of type S-I to clump but do not affect type S-II; the reverse is true as well.

A further point must be made relative to these disease organisms; that is, all types of S forms can mutate to R ones, and vice versa. The mutational changes from R to S always restore the same type (I, II, III, or IV) that gave rise to the R form in the first place. The changes can be represented as follows:

$$S\text{-I} \rightarrow R \rightarrow S\text{-I} \rightarrow R \rightarrow S\text{-I} \ldots$$
$$S\text{-II} \rightarrow R \rightarrow S\text{-II} \rightarrow R \rightarrow S\text{-II} \ldots$$
$$S\text{-III} \rightarrow R \rightarrow S\text{-III} \rightarrow R \rightarrow S\text{-III} \ldots$$
$$S\text{-IV} \rightarrow R \rightarrow S\text{-IV} \rightarrow R \rightarrow S\text{-IV} \ldots$$

The rough forms exhibit none of the types indicated by the Roman numerals because they have no polysaccharide coats; the coats are the antigens that evoke the corresponding antibodies. Nevertheless, each R form "remembers" its roots, and when it reverts by mutation to an S form, it invariably reverts to the S form from which it was derived. Thus, R(I), R(II), R(III), and R(IV) indicate the now-unexpressed coat type of the rough forms.

Frederick Griffith studied pneumococcus during the 1920s. Among other studies were several in which he inoculated mice with living R forms, heat-killed S forms, or both living R and heat-killed S forms. As expected, the first two treatments had no noticeable effect on the inoculated mice: rough bacteria are not pathogenic, dead S forms do not produce infections. When the two forms of pneumococcus were inoculated into the mice simultaneously, however, the animals frequently died of pneumococcal infection. From these (dead) mice, living S pneumococci could be obtained and cultured. These S forms, unlike those in the tabulation above, were of the type of the *dead* S forms used in the original inoculation; they were not mere reversions of the R individuals to their ancestral type:

$$\text{(Dead) S-II} + R(I) \rightarrow \text{(Live) S-II}$$

Once mystical influences are brushed aside (including the possibility that any material produced by the living rough individuals might resurrect the dead ones whose proteins were coagulated by exposure to heat), the inescapable conclusion is that some substance produced by the dead S-II individuals has altered the living rough ones so that they have assumed a different serological type.

But what substance? Experiments are difficult to carry out in mice; workers had to transfer the system to test tubes and petri dishes to perform the necessary chemical fractionation of dead S individuals that allowed the systematic testing of each fraction in the absence of others. What substance isolated from dead S-II cells converts live R(I) cells into live S-II cells?

The answer to this question was obtained in 1944 by three members of the Rockefeller Institute: O. T. Avery, C. M. MacLeod, and M. McCarty. Today, it is easy to oversimplify what was then an exceedingly complex task in organic chemistry. It is also easy to overlook the consternation of the three scientists as they carried out their tests. Time and time again they ran through the procedures for isolating the main constituents of pneumococci: the polysaccharide coat, the proteins, and a substance possessing the characteristics of DNA. Repeatedly, they found that the polysaccharide had no effect on rough cells; none were transformed into smooth cells. Repeated tests of the proteins yielded negative results as well. Each test that utilized the DNA fraction gave positive results; smooth colonies appeared here and there on the petri dishes when the treated rough cells were spread on the agar plates for testing. Not only were there smooth colonies, but the type (S-I, S-II, etc.) corresponded to the type of smooth cells that provided the compounds being tested, not to the type that was latent within the R cells. DNA obtained from S-II cells, that is, transforms R(I) cells into S-II

Further tests were made of the presumed DNA transforming material. It was exposed to a series of proteolytic enzymes (digestive enzymes that degrade proteins); the transforming ability remained. The transforming material was exposed to ribonucleases, enzymes capable of degrading RNA; still the transforming ability remained. The material, highly purified and extremely efficient at transforming rough cells, was exposed to an enzyme capable of degrading DNA into small fragments. Now the transforming ability vanished. The degraded DNA of S cells no longer transformed rough cells into smooth ones.

Shortly after leaving the U.S. Air Force in 1946, I returned to Columbia University to resume life as a graduate student. The Dobzhanskys invited the (newly married) Wallaces and the Mirskys to a Russian Easter lunch. After lunch I remained at the dining room table with Dr. Mirsky while the others moved to the living room. There were two papers that I must read in order to catch up on my three-and-a-half-year absence, Mirsky told me: Avery, MacLeod, and McCarty's paper on transformation in pneumococcus and Sonneborn's analysis of kappa particles in paramecia, an apparent example (shown later to be a symbiont) of cytoplasmic inheritance. Imagine! Only two papers to correct a three-and-a-half-year absence. Today, each Ph.D. candidate is expected to have published at least two papers in order to have a realistical shot at a postdoctoral fellowship.

The data on transformation did not convince many geneticists and biochemists that DNA is genetic material. Mirsky argued that traces of protein too small to be detected even by those as cautious as Avery and his colleagues might be responsible; this argument could be countered only by a superhuman effort at purification by precipitating the DNA, resuspending it in solution, and reprecipitating it once more over and over again. This repetitive procedure might "wash out" a casual contaminant, but it need not eliminate a protein that reacted with and attached itself to the DNA.

Directed mutation was the term many geneticists used in discussing bacterial transformation. In their textbook *General Genetics*, A. M. Srb and R. D. Owen (1952, p. 250) reported that the "actual basis for the activity of transforming principles [which, they said, appear to be a nucleic acid] is, however, unknown." In another popular genetics textbook, E. W. Sinnott, L. C. Dunn, and Th. Dobzhansky (1958, p. 240) referred to transformation as "directed mutation" in contrast to the random, undirected mutations induced by mutagenic radiation. The following quotation reveals how far DNA was from the authors' concept of the gene: "Now, this nucleic acid is continuously produced in the body of a given type of pneumococcus, and it is able under certain conditions to become established in cells derived from other types and subsequently to go on producing more of itself." C. D. Darlington and K. Mather (1949) took a metabolic approach: "Avery and others found that if the capsule is lost, the specificity of a strain A can be transferred to a strain B, by feeding it with dead cells of A, or, even more precisely, by feeding it with [deoxyribose] nucleic acid from the capsules of A." The misinformation in this sentence makes the prevailing silence of many books of the era welcome.

To anticipate what I will demonstrate in the next section, I shall merely say here that the common error of those discussing Avery's work was to imagine that DNA *influenced* or *modified* genes (hence, "*directed* mutations") rather than that DNA *was* the genetic material itself (see Figure 7-1).

Bacteriophage, their coats, and their DNA

The search for the gene now enters the realm of viruses; some of these are known as bacteriophage, that is, "organisms that eat bacteria." The reader may have noticed that the experimental organisms used by geneti-

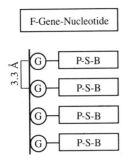

Figure 7-1. The structure of the gene as envisioned by Darlington and Mather (1949); subsequently, it proved to be quite erroneous. The protein fiber that constitutes the chromosome is designated F. Attached to it are the genes (G), and attached to each gene is a nucleotide consisting of phosphoric acid (P), deoxyribose (S), and a nitrogenous base (B). That the diagram includes exactly four genes suggests that the four bases (adenine, thymine, guanine, and cytosine) were thought to occur repetitively in agreement with the prevailing (Levene's) tetranucleotide model (see p. 128).

cists have become ever smaller as the search has progressed. Castle worked primarily with rats and mice, but he took credit for introducing *Drosophila melanogaster* (then known by another scientific name) to Thomas Hunt Morgan. Beadle began with corn at Cornell University but switched to the smaller fruit fly, only to leave it for *Neurospora.* A number of confirmed *Drosophila* workers left their tiny flies for *E. coli,* the colon bacillus. Now we turn to a very small, tadpole-shaped organism that has the capacity to infect bacteria and, after a lapse of *minutes,* cause each infected bacterium to burst, releasing scores of new phage particles. In contemplating the steady decrease in the size of the experimental organism, I am reminded of the former heavyweight boxing champion Joe Louis, who, when asked how he intended to handle an opponent noted for his elusiveness in the ring, merely said, "He can run but he can't hide." Geneticists have over time given the gene less and less space in which to hide.

Bacteriophage (or, to shorten the name, *phage* or *phage particles*) are remarkably simple organisms. Externally, they resemble robot tadpoles that are equipped to land on a very small moon (Figure 7-2). Although once believed to "swim" toward the bacterium head-first (Figure 7-3), improved electron micrography has revealed that they actually attach to the surface of the bacterium tail-first. I nearly wrote "attach themselves," which, for a mere collection of protein and DNA molecules, would have exaggerated the willfulness of phage infection. The "tail" fibers interact with certain proteins on the bacterial surface; the result is infection, lysis, and death of the bacterial cell. A few bacterial cells are resistant to such infection. This resistance can be ascribed to mutant forms of the bacterial surface proteins; the phage's tail fibers do not match the mutant form.

Phage geneticists have learned a great deal about the morphology and structure of phage particles, the kinetics of infection, the enzymatic and

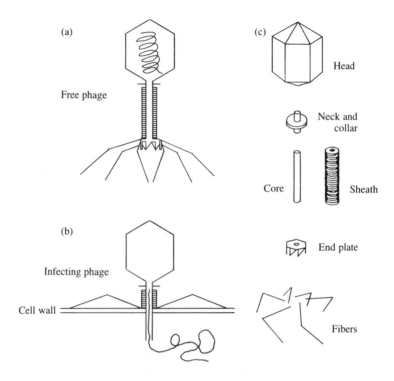

Figure 7-2. The construction of a bacteriophage. (a) The protein head (at top) contains the phage's DNA, tightly coiled and packed. The fibers attach to protein receptors on the cell wall of the bacterium. (b) DNA from the phage is injected into the bacterium, where it directs the synthesis of viral enzymes and "coat" proteins. (c) Many of the parts illustrated in the exploded view of the phage particle are self-assembling—joining the different pieces requires neither enzymes nor energy. (Redrawn from R. S. Edgar and R. H. Epstein, "The Genetics of a Bacterial Virus." Copyright © 1965 by Scientific American, Inc. All rights reserved.)

other alterations that occur within infected bacterial cells, and the assembly of new phage particles. By using various means, including chemical solutions, to break open bacterial cells at different times after infection, these workers found that for 10 minutes no mature phage can be found within the bacterium (Figure 7-4), but after a lapse of several more minutes, the number of mature phage (those capable of reinfecting bacteria) steadily and rapidly increases. Normally, this increase occurs only after 20 minutes has elapsed; rupturing the bacterial cells prematurely reveals that the mature particles accumulate within the bacterial cell before it finally bursts. The

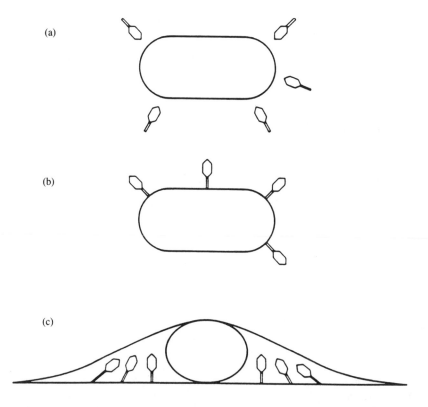

Figure 7-3. The "attraction at a distance" seemingly exhibited by bacteriophage (a) implied that they swam, spermlike, toward a nearby bacterium. Improved mounting techniques that removed surface tension as a distorting force revealed that phage particles attach to bacteria by their tails, not by their heads (b). The "attraction" was then seen (c) to be an artifact caused by the surface tension of the evaporating mounting liquid; phage particles, which were standing tail-down, were toppled head-first toward any nearby bacterial cell. (Modified from Wallace and Simmons, 1987.)

work described here so briefly was carried out largely by a group of young biologists working in collaboration with Max Delbrück, Salvadore Luria, and A. D. Hershey, three scientists who shared the Nobel Prize in physiology and medicine in 1969.

Where in this intricate process of infection and reproduction was one to find the gene? The answer was provided by A. D. Hershey and his technician-collaborator, Martha Chase. (Hershey later said that only Chase had the power of concentration needed to follow the intricate protocol

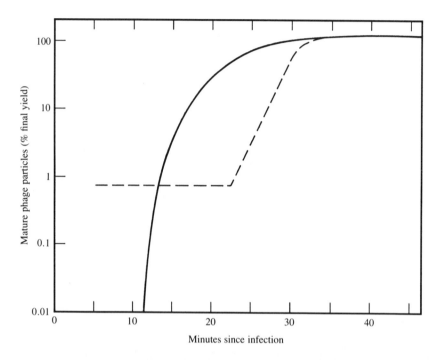

Figure 7-4. An early demonstration that mature phage particles (those capable of infecting bacteria) normally form within the host and increase in number before the bacterial wall ruptures and releases them. The normal course of events is represented by the dashed lines. About 20 minutes after infection a sudden (100-fold) rise in the number of phage particles reveals that infected bacteria are rupturing. If the bacteria are experimentally ruptured prematurely (solid line), it is apparent that the number of mature phage particles starts increasing about 12 minutes after the original infection. The first 10–12 minutes during which no mature phage can be detected is known as the eclipse period. (Redrawn from Stent, 1963, by permission of W. H. Freeman.)

demanded by these experiments.) Phage particles had been analyzed chemically, and it was known that they consist of an outer protein jacket and an inner core of DNA. Hershey and Chase had available to them isotopes of sulfur (^{35}S) and phosphorus (^{32}P); these and other radioactive elements became readily available for biological research following the end of World War II.

Bacteria grown on media containing ^{35}S incorporate this isotope into their proteins; two amino acids—cysteine and methionine—contain atoms of sulfur. If these sulfur-fed bacteria are infected with bacteriophage, the progeny phage particles that are produced contain ^{35}S atoms in their protein jackets.

Bacteria grown on media containing ^{32}P incorporate these atoms into their DNA. There is no phosphorus in proteins, nor is there sulfur in DNA. If bacteria growing in a ^{32}P-containing medium are infected with bacteriophage, the phage particles produced contain DNA that is also labeled with radioactive ^{32}P.

In an extensive series of experiments involving radioactively labeled bacteriophage, Hershey and Chase (1952) studied the roles of the protein and DNA components of phage particles; they accomplished this by determining the fate of ^{35}S and ^{32}P under a number of circumstances. An early experiment had revealed that ^{35}S (and hence the phage protein) remains on the outside of the bacterium, whereas the DNA is found within the bacterial cell. When cellular debris rather than intact cells was "infected" with labeled phage, the phage protein became and remained attached to the bacterial debris; the ^{32}P (hence the DNA), in contrast, was to be found in solution. (We can anticipate the ultimate interpretation [see Figure 7-2b] by suggesting that the DNA was injected through the fragment of bacterial wall—however, in this experiment there was no bacterium on the other side!) Finally, no ^{35}S reappeared in progeny phage particles following infection by labeled parental phage; in contrast, 30% or more of the ^{32}P of labeled parental phage reappeared within the DNA of progeny phage.

Today, these experiments are often accepted as proof that the gene is DNA, or that DNA is the genetic material. Hershey and Chase were not that sanguine. As F. H. Portugal and J. S. Cohen (1977, p. 183) say: "The results of the experiments were, in fact, equivocal. For example, more than 20 percent of the nucleic acid isotope [^{32}P] was released from the bacteria while approximately the same amount of protein isotope [^{35}S] remained attached. If the same arguments used against the experiments of Avery and his co-workers . . . were applied to these experiments, then clear conclusions could not be drawn."

Nevertheless, Hershey and Chase are often credited with having performed *the* experiment that finally located the gene, thus ending a long search. The phage particle is like a syringe. After attaching to the bacterium, the DNA core is injected into the bacterial cell while the protein jacket remains outside. The DNA carries all the information needed not only to direct its own replication but also to direct the synthesis of bacteriophage coat proteins, proteins quite unlike those normally made by a bacterial cell.

Why erect such a consistent story on the basis of an experiment that, even in the view of its authors, was inconclusive? In part, the experiment represented the "final straw." Its ambiguities could be understood (the Waring blender that was used to dislodge infecting phage from bacterial cells is not, after all, the most delicate of hi-tech equipment) and thus could be explained away. An important factor was, however, the psychology of scientists. The Phage Group (as the young scientists surrounding Delbrück, Luria, and Hershey called themselves) had introduced a new organism into genetic research and had developed an entirely new set of procedures, techniques, and ways of thinking. When the dust finally settled and they were convinced that DNA and the gene are synonymous (if not identical), their search for "historical" beginnings went no further than Hershey and Chase. They were not interested in pursuing the search into a wilderness (the one that we have traversed) in which they may have been uncomfortable at best, and perhaps even intellectually bewildered. Only a small number of this group, for example, would have anticipated Muller's ancient assertion that the hereditary material must be able to direct its own replication, to change, and then to direct the replication of the changed form as readily as it had the original form. Few people realize that if DNA had not conformed to Muller's early stipulations, the search for the gene would still be in progress.

Box 7-1. Isotopes

How many of my readers remember the periodic table (Figure 7-5) from high school or college chemistry classes? In it are the chemical elements from the lightest (hydrogen) to the heaviest, listed in order of their atomic numbers (H, 1; He, 2; O, 8; etc.). Also shown are their atomic weights, numbers that, for light elements other than hydrogen, are approximately twice the atomic number. Thus, the atomic weight of neon (Ne) is 20.183; the atomic number is 10.

I mention neon because it was the first element whose atomic mass was determined. This was done by passing positive ions of neon through a magnetic field and measuring the deflection of their paths from the path followed when the magnetic field was removed. The observed deflection, which should have indicated an atomic weight of 20.183, indicated instead 19.999—too far off to be caused by experimental error. Close examination of the fluorescent screen that revealed the impact of the

Figure 7-5. The periodic table of the chemical elements. Shown for each element are the atomic number (top) and atomic weight (bottom). Isotopes of many of these elements have been extremely useful in biological (including genetic) research. Not shown in this figure are the rare earths (atomic numbers 58–71).

																	0
1 H 1.0																	2 He 4.0
3 Li 6.9	4 Be 9.0											5 B 10.8	6 C 12.0	7 N 14.0	8 O 16.0	9 F 19.0	10 Ne 20.2
11 Na 23.0	12 Mg 24.3											13 Al 27.0	14 Si 28.1	15 P 31.0	16 S 32.1	17 Cl 35.5	18 Ar 39.9
19 K 39.1	20 Ca 40.1	21 Sc 45.0	22 Ti 47.9	23 V 50.9	24 Cr 52.0	25 Mn 54.9	26 Fe 55.8	27 Co 58.9	28 Ni 58.7	29 Cu 63.5	30 Zn 65.4	31 Ga 69.7	32 Ge 72.6	33 As 74.9	34 Se 79.0	35 Br 79.9	36 Kr 83.8
37 Rb 85.5	38 Sr 87.6	39 Y 88.9	40 Zr 91.2	41 Nb 92.9	42 Mo 95.9	43 Tc (97)	44 Ru 101.0	45 Rh 102.9	46 Pd 106.4	47 Ag 107.9	48 Cd 112.4	49 In 114.8	50 Sn 118.7	51 Sb 121.8	52 Te 127.6	53 I 126.9	54 Xe 131.3
55 Cs 132.9	56 Ba 137.3	57 La 138.9	72 Hf 178.5	73 Ta 180.9	74 W 183.9	75 Re 186.2	76 Os 190.2	77 Ir 192.2	78 Pt 195.1	79 Au 197.0	80 Hg 200.6	81 Tl 204.4	82 Pb 207.2	83 Bi 209.0	84 Po (210)	85 At (210)	86 Rn (222)
87 Fr (223)	88 Ra (226)	89 Ac (227)	90 Th 232.0	91 Pa (231)	92 U 238.0	93 Np (237)	94 Pu (242)	95 Am (243)	96 Cm (247)	97 Bk (247)	98 Cf (249)	99 Es (254)	100 Fm (253)	101 Md (256)	102 No —	103 Lr —	

neon ions showed two additional, much fainter points of impact. Further calculation revealed that these overlooked neon ions had atomic weights of 21.998 and 20.999. Rather than consisting solely of atoms with an atomic weight of 20.183, neon consists instead of a mixture of different atoms. Most of these have an atomic weight of 20, but other, heavier ones have weights of 21 and 22: ^{20}Ne, ^{21}Ne, and ^{22}Ne. The different types of atoms are called *isotopes* of neon. Hydrogen is also a mixture of isotopes: ^{1}H, deuterium (^{2}H), and tritium (^{3}H). The average atomic weight of this mixture is 1.008.

Isotopes are exceedingly useful in biological research. In some instances (as we shall see in the next chapter) the difference in mass is useful; in other instances the instability of the atomic nucleus and the radiation emitted as the less stable isotopes decay is useful. Most useful of all is the identical behavior all isotopes exhibit in chemical reactions: ^{14}C behaves just like the more common ^{12}C, ^{15}N like ^{14}N, ^{32}P like ^{31}P, and ^{35}S like ^{32}S. Hershey and Chase labeled DNA with ^{32}P (which has a half-life of 14.3 days; i.e., only half of the original number of ^{32}P atoms remain after 14.3 days). Each ^{32}P atom changes to a normal sulfur atom, ^{32}S, when it decays. Sulfur, of course, is not a constituent of DNA; therefore, each disintegration of ^{32}P in labeled DNA causes a break in the DNA molecule. In an experiment that lasts only hours, results are not greatly affected by half-lives measured in days (that of ^{35}S is 88 days). It is interesting to note that the half-life of ^{14}C is 5730 years. Half of the ^{14}C atoms that were present in the world when the oldest Egyptian pyramid (the "step pyramid") was built are still present today; one quarter will remain at an equally distant time in the future.

Box 7-2. Viruses

Viruses, including bacteriophage, are extremely small, much smaller than bacteria. Many of them cause disease, and for that reason have been known by their effects for ages. It was only about 1900, however, that M. W. Beijerinck decided to isolate the organism that causes tobacco mosaic disease from the sap of sick tobacco plants. After passing the sap through an extremely fine filter, one known to be fine enough to catch bacteria, he found no bacteria. He did show, however, that the filtered sap put on a healthy plant could cause the mosaic disease. Thus the infective agent was known to be smaller than the pores of a bacterial filter. The "agent" consisted of discrete particles, however; that fact was revealed by diluting the filtrate to different degrees and smearing some of each dilution on tobacco leaves that had been lightly scraped with sandpaper: the number of necrotic spots on the injured leaves decreased in direct proportion to the degree of dilution of the filtered sap.

Wendell Stanley, a chemist, caused a sensation about 1930 when he obtained crystals of tobacco mosaic virus (TMV) in his laboratory. Until that time, people were content to think of TMV and other viruses as tiny living things. Crystals, however, are the chemist's proof that a substance—a chemical substance—has been purified. The TMV crystals, incidentally, were capable of causing mosaic disease when redissolved and painted on leaves of healthy tobacco plants.

A slight digression may be in order at this point. Human beings have the capacity to establish verbal and conceptual categories when corresponding categories do not exist in the real world. Hence, we speak of "drugs" and "nondrugs" while discussing the possible legalization or decriminalization of certain substances. The world consists of chemical molecules that range from amino acids, carbohydrates, and vitamins (essential nutrients) through caffeine, aspirin, and alcohol (elective compounds) to mind-boggling substances. "Drugs" is not a natural category in the real world. Neither is "carcinogen"; chemicals do not come in the form of cancer-causing (carcinogenic) and non-cancer-causing substances. There are many kinds of cancer (which itself is a man-made concept) that arise from many different chemical treatments. This fact, however, does not inhibit the many tests that are needed to determine whether a given substance is or is not carcinogenic. *Life* is a term derived from the convenient but artificial division of objects into living and nonliving. Earlier I emphasized the contrast between an old textbook's use of "living protoplasm" and Watson's assertion that there are no special chemicals or chemical reactions in cells that could not, with effort, be synthesized or carried out in the laboratory. This contrast looms greatest when one considers viruses, small bundles of proteins and nucleic acids that can either direct their own synthesis within living cells or form crystals when purified in a chemical laboratory.

In addition to plant viruses, there are a host of animal viruses responsible for diseases such as herpes, smallpox, polio, 24-hour intestinal upsets, and AIDS. These viruses live within the host's cells and as a rule are shed individually by these cells, rather than by the complete disruption of each infected host cell. The sinister aspect of the AIDS-causing virus is its penchant for entering and altering precisely those cells that normally protect the host against disease-causing organisms. Thus, as matters now stand, the AIDS virus virtually guarantees its host's demise; this situation can be tolerated by a disease-causing organism only if the infection of new hosts is sufficiently rapid to maintain the disease in the population of hosts. If the incubation period is eight to ten years, as it is in the case of AIDS, and if the virus is transmitted primarily by sexual contact (neglecting hypodermic needles), either monogamy, continuous testing of all sexually active persons for evidence of infection, or an extremely efficient barrier to infection will be required to control the disease by "ecological" (i.e., nonmedical) means.

Two persons can be credited with the discovery of bacteriophage: F. W. Twort and Félix d'Hérelle. Both noted that thriving bacterial colonies or cultures can be destroyed (become clear rather than turbid in the case of liquid cultures) overnight by the presence of even a single phage particle. If an infected bacterium releases 100 new phage particles in 20 minutes, and if each of these infects a bacterium and releases 100 more in another 20 minutes (10^4 particles by now), it is obvious that relatively few rounds of infection would be needed to infect and destroy a culture tube containing even 10^{12} bacteria.

According to G. S. Stent (1963), F. M. Burnet was the pioneer who devised many of the basic procedures in the study of bacteriophage in about 1930. Why, we might ask in conclusion, has bacteriophage been so prominent in our search for the gene when plant viruses were under investigation earlier—even crystallized by the time Burnet was publishing his results? The answer, I believe, is that bacteria are free-living, single-celled organisms. Those working with both plant and animal viruses and with bacteriophage took pains (as did the plant hybridizers of the nineteenth century in studying the role of pollen in fertilization) to show that a single viral particle was capable of infecting a single host cell. In the case of plants and higher animals, however, these single cells are parts of organized tissues and whole individuals. This fact imposes a severe restraint upon experimental techniques (and lines of reasoning) that might be employed in future research. Bacteria, on the other hand, exist as individuals; phage particles also exist as individuals. Thus, a whole battery of experiments involving different ratios of phage particles to bacterial cells can be carried out. In each of these, individual bacteria can be isolated and studied. The speed with which science advances depends upon the range of questions that practicing scientists are able to ask. The Phage Group had an opportunity to ask many more questions than did other virologists.

Box 7-3. The Cold Spring Harbor Laboratories

The Phage Group, of whom Delbrück, Luria, and Hershey were the acknowledged leaders, met during the summer months at Cold Spring Harbor, Long Island, New York. The town itself is an old whaling village situated at the tip of one of Long Island Sound's numerous inlets (harbors), between the towns of Huntington and Oyster Bay. The first laboratory built there was supported by wealthy local residents; it was called simply The Biological Laboratory of the Long Island Biological Association. Because the Biological Laboratory was there, the Carnegie Institution of Washington was persuaded to create a Department for the Study of Experimental Evolution on adjacent land; later this department was renamed the Department of Genetics.

Whatever the quality of the research done at Cold Spring Harbor (and it tended to be good), Reginald Harris, director of the Biological Laboratory, showed a stroke of genius when he initiated the Cold Spring Harbor Symposia on Quantitative Biology in 1933. As he said in introductory remarks for the symposium volume of the following year, "Cold Spring Harbor Symposia on Quantitative Biology are an experiment in scientific procedure, a natural outgrowth of the Biological Laboratory's policy of fostering a closer relationship between biology and the basic sciences." How well that experiment has succeeded (and, perhaps, how the attitudes of scientists have changed over nearly sixty years) can be judged from the designation (by the *New York Times*) of the Cold Spring Harbor Symposium as one of several meetings for which an invitation to participate is regarded as a symbol of prestige by the scientific community.

During the 1950s the laboratory was exceedingly informal and low-key. Because bacteriophage had only recently been adopted as an organism for serious genetic study, a course in phage techniques was offered each summer. During the early years of this course, those accepted as students were university professors who, having learned the details of phage manipulations, would in turn teach graduate students at their home institutions. Many of the students at that time were physicists (as Max Delbrück had been) who were accustomed to mathematical reasoning, attracted by the simplicity of phages and phage reproduction, and discouraged by the official secrecy that had enveloped atomic research. One of the more colorful students who took the course and returned to Cold Spring Harbor at frequent intervals was Leo Szilard, the physicist who encouraged Albert Einstein to write President Franklin D. Roosevelt urging the construction of the atom bomb and who actually held a patent on the chain reaction that is at the heart of nuclear power.

Hershey joined the Department of Genetics in 1950. Already present were M. Demerec, director of both the Department of Genetics and the Biological Laboratory; Barbara McClintock, an outstanding corn geneticist; B. P. Kaufmann, a cytogeneticist; and Carlton McDowell, a mouse geneticist. The staff at the Biological Laboratory were younger, less well established persons working exclusively on "soft" money (i.e., supported by grants or contracts with federal agencies).

Barbara McClintock deserves special mention; she was awarded the Nobel Prize in physiology and medicine in 1983 for her work on transposable elements in corn. In the course of her studies of abnormal chromosomal breakage and re-fusions, McClintock encountered and concentrated her attention on kernels that possessed a mottled appearance (similar kernels can be seen on many ears of so-called Indian corn available in supermarkets near Halloween). The mottling she studied was the result of a complex system involving at least three components, but, because of her familiarity with the experimental organism, she succeeded in resolving the mysteries that confronted her. One facet of her analysis revealed that genes could move from chromosome to chromosome, at times in a nearly predictable manner. McClintock's system was investigated at the molecular level only after similar effects had been observed and analyzed in bacteria. In retrospect, the contrast between Hershey, who was zeroing in on the resolution of DNA's role in phage reproduction, and McClintock's beginning on a problem that seemed so hopelessly complex that no solution would be forthcoming is unbelievable. Both, however, solved the problems they set out to solve; both have been recognized for the excellent research they carried out—both are Nobel laureates.

Box 7-4. Knowing one's organism

The success of biological research rests largely on the astute choice of experimental material. A few altruistic individuals choose organisms and problems that cannot possibly be solved within their lifetimes; they are to be saluted by the rest of us. The efforts of Charles Burnham to obtain blight-resistant American chestnut trees by artificial selection is an example of this selflessness. In all probability, Dr. Burnham will never see the outcome of his labor.

Virtually all my readers will recall from either high school or college biology the advantages that led to genetic research on the fruit fly, *Drosophila melanogaster:* ease of culturing, short generation time, and many progeny (often hundreds) from single-pair matings.

Barbara McClintock is a maize geneticist. Her graduate training was at Cornell University, largely under the direction of R. A. Emerson, head of the Department of Plant Breeding and the "overseer" of a group of young botanical geneticists (Figure 5-8) that also included Charles Burnham, George Beadle, Marcus Rhoades, and Milislav Demerec. Dr. McClintock's precision as a member of that group is legendary; a flash flood, for example, destroyed everyone's research one summer except McClintock's, whose plants were so carefully labeled (and the map of her experimental plot so carefully prepared) that each one could be returned to its original hill.

While she was a member at the Department of Genetics of the Carnegie Institution of Washington at Cold Spring Harbor, New York, Dr. McClintock learned of the entrance of Max Delbrück (a physicist) and Salvadore Luria (a medical doctor) into biology by way of bacteriophage. Her reputed comment was that they would not succeed because they did not know their organism. Barbara McClintock *knows* corn! She loves her corn plants, and she knows every detail about their development, growth, and reproduction. When tapioca could not be obtained from Southeast Asia during World War II, for example, Barbara McClintock successfully engineered (by genetic crosses, not by molecular techniques) a strain of corn that could be used in the manufacture of glue, the main commercial product of imported tapioca.

In the case of the Phage Group, knowing their organism and answering the questions they asked were nearly identical matters: that phage DNA is injected into bacteria during infection while the protein coat remains outside more or less coincides with the knowledge that a bacteriophage consists of a genetically "inert" protein coat and a DNA core.

One cannot claim that the members of the Morgan group "knew" their organism, the fruit fly. When Calvin Bridges encountered a fly with two pairs of wings rather than the single pair characteristic of the order Diptera (*di* = twice + *pteron* = wing), he concluded that it had developed a second thorax; hence the designation *bithorax*. Bridges did not know the embryological origin of the insect thorax, which arises from three embryonic segments—prothorax, mesothorax, and metathorax. The mesothoracic segment in two-winged flies bears the flying wings. The metathoracic segment bears the club-shaped halteres, modified wings that act as gyroscopic sensing organs during flight. The mutant *bithorax* undergoes an abnormal development in which the halteres of the metathorax become winglike.

The admonition "know your organism" was taken seriously by Sidney Brenner and his associates in developing a small nematode, *Caenorhabditis elegans,* into an experimental organism. In seeking an organism that might yield information about both development and brain function, Brenner chose this nematode because it has nerves, muscles, an intestine, it reproduces—and, if you hit it, it reacts (Roberts, 1990). During its development this nematode comes to possess 959 somatic cells—no more, no less. The entire developmental pedigree of every somatic cell is now known.

The nervous system of *C. elegans* comprises 302 neurons. All connections (the "wiring" diagram) of these neurons are also known. This knowledge was gained only by assembling thousands of serial electron micrographs—acres of prints—and carefully noting the course of each neuron and its connections with other cells, neurons or otherwise.

The next research effort involving Brenner's small nematode is to sequence its entire genome. By knowing the origin and fate of every cell, by knowing every connection in its nervous system, *and* now by determining every one of its 100 million base pairs, those working on *C. elegans* hope to make it the standard for all future genomic studies, including the Human Genome Project.

8   The Double Helix

The gene, for our purposes, has now been shown to be DNA. The end of this portion of the search was not as spectacular as it might have been; we more or less stumbled onward from UV action spectra to Avery's transformation to Hershey's experiments with ^{35}S- and ^{32}P-labeled bacteriophage—but nevertheless, the cumulative evidence has been compelling. Now, however, matters become somewhat inverted. What is it about DNA that makes it suitable to serve as genetic material—as the gene?

Our last detailed look at DNA was in Chapter 6, where we considered the chemical composition of DNA. As chemists struggled to unravel the manner in which the deoxyribose molecule, the phosphoric acid, and the purine or pyrimidine base were attached to one another, various workers broke rank and allowed themselves speculative guesses. One guess was that DNA is a circular compound containing four sugar molecules, one nitrogenous base of each sort (adenine, thymine, guanine, and cytosine), and four molecules of phosphoric acid. The entire molecule would look much like a quartet of performers at a barn dance (Figure 8-1).

Linear models were more popular. In 1935 P. A. Levene had proposed a tetranucleotide structure for the DNA molecule with purines and pyrimidines following one another along its length in lockstep order (Figure 8-2). Oddly, the term *tetranucleotide* had been used by Albrecht Kossel before the turn of the century in proposing four kinds of filamentous DNA molecules, each containing only one of the four nitrogenous bases.

Describing the correct solution is not made easier by DNA's being a term encountered daily in the local paper, by James D. Watson's (who with Francis Crick arrived at the correct solution) best-selling book about his

128

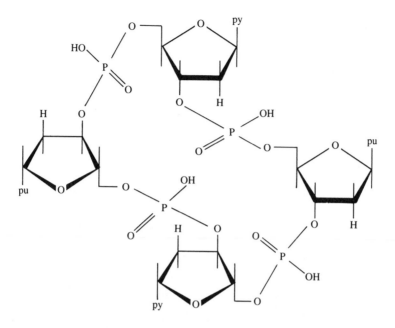

Figure 8-1. In attempting to determine the structure of complex molecules, chemists construct models consistent with the available observed data. This model represents a proposed circular molecule of deoxyribonucleic acid containing deoxyribose, phosphoric acid, and the four nitrogenous bases. It fails to account for the long, filamentous structure of isolated DNA molecules. (pu = purine; py = pyrimidine.)

role in the fascinating process of creating the correct model, and by the occasional television movie in which actors play out a glamorized version of what transpired during the early 1950s at Cambridge University where Watson and Crick worked.

Levene's tetranucleotide model required that any sample of DNA subjected to degradation and chemical analysis should contain the four bases in equal quantities. The data relevant to this point may, in the 1950s, have been obscure. Mirsky (1951, p. 135), for example, cited data of the following sort:

Source	A	G	T	C	Total recovered
Calf thymus	27.6%	23.5%	28.0%	19.7%	98.8%
Wheat germ	27.4%	23.6%	27.0%	16.4%	94.4%

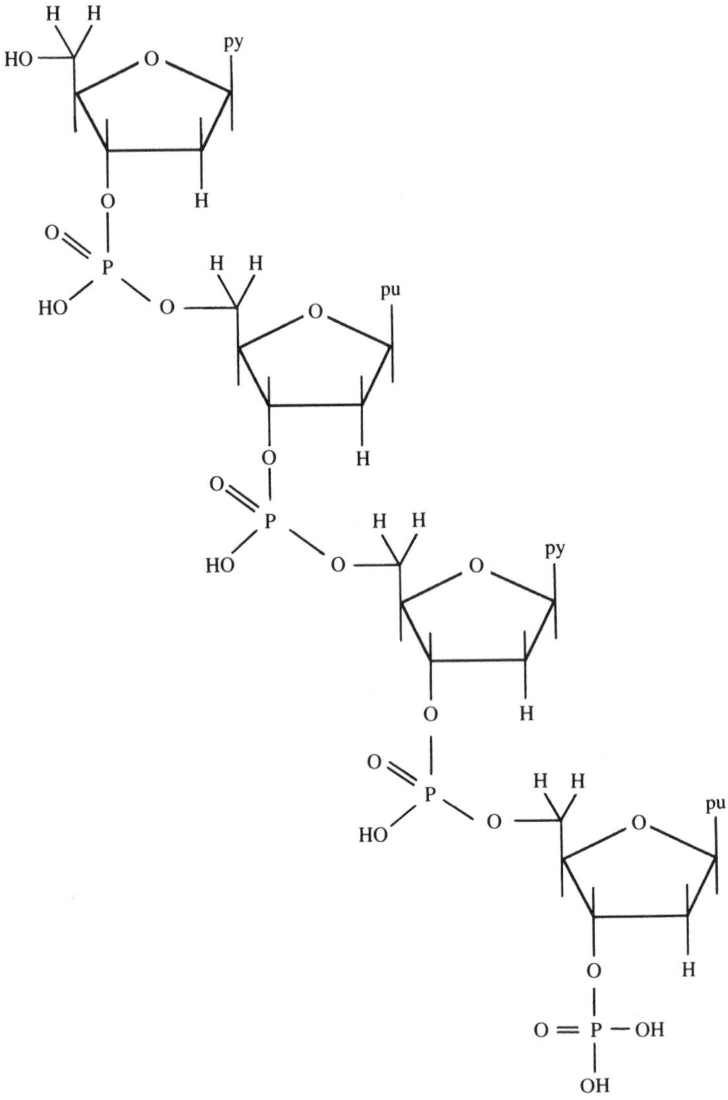

Figure 8-2. The tetranucleotide structure of DNA proposed by Levene in 1935. This structure could, of course, be repeated indefinitely, thus forming long filaments. As postulated by Levene, however, each repeating unit of the sort shown here contains one purine and one pyrimidine molecule of each type: guanine, adenine, cytosine, and thymine. Thus, the expected ratio of G : A : C : T would be 1 : 1 : 1 : 1, an expectation that was disproved by Erwin Chargaff in 1950.

Table 8-1. A retabulation of data presented by Chargaff (1950, p. 205) but emphasizing the A : T and G : C ratios he had suggested were close to 1.00. Although the G : C ratio departs considerably from 1.00 (a ratio that Chargaff had no reason to expect), Chargaff did mention earlier (p. 204) that "the extremely robust cleavage methods with mineral acids usually employed must have led to a very considerable degradation of cytosine to uracil." In brief, the calculated G : C ratios, because of the loss of cytosine, should be larger than 1.00, as indeed they are.

Organism and tissue	A : T	G : C	% recovery
Ox thymus	1.13	1.31	92
Ox spleen	1.06	1.28	86
Human sperm	0.92	0.97	94
Human thymus	1.00	1.19	91
Human liver carcinoma	1.00	1.20	87
Yeast	1.00	1.14	84
Avian bacillus	1.09	1.08	77

These are not exactly the expected equal proportions, but the deviations from equality might not deter a true believer. Only sentences before citing these data, Mirsky had pointed out that "it is well known that in the course of hydrolysis *different* amounts of purine and pyrimidine bases may be destroyed" (p. 134). A similar point must have been on Watson's mind when he referred to a colleague's assurance that if Erwin Chargaff (a DNA chemist) said that the amount of guanine equaled that of cytosine, then it probably did not (see Table 8-1). That sort of argument made Watson suspicious of data concerning base ratios.

The demise of the tetranucleotide hypothesis (which called for equal quantities of the four purine and pyrimidine bases), if one accepted the data, lay in Chargaff's (1950) analyses of DNA from a variety of sources. In the ox, for example, Chargaff found the adenine-to-guanine ratio to be 1.29, and that of thymine to cytosine to be 1.43. The same values for human DNA were 1.56 and 1.75; for yeast, 1.72 and 1.90; but for an avian bacillus (*Mycobacterium avium*), 0.4 and 0.4. These values are far from equality, and they led Chargaff to state, "The results serve to disprove the tetra-nucleotide hypothesis" (p. 206). He went on to say (note carefully the qualifying phrase): "It is, however, noteworthy—whether this is more than accidental, cannot yet be said—that in all [deoxypentose] nucleic acids examined thus far the molar ratios of total purines to total pyrimidines, and also of adenine to thymine and of guanine to cytosine, were not far from 1 " (p. 206).

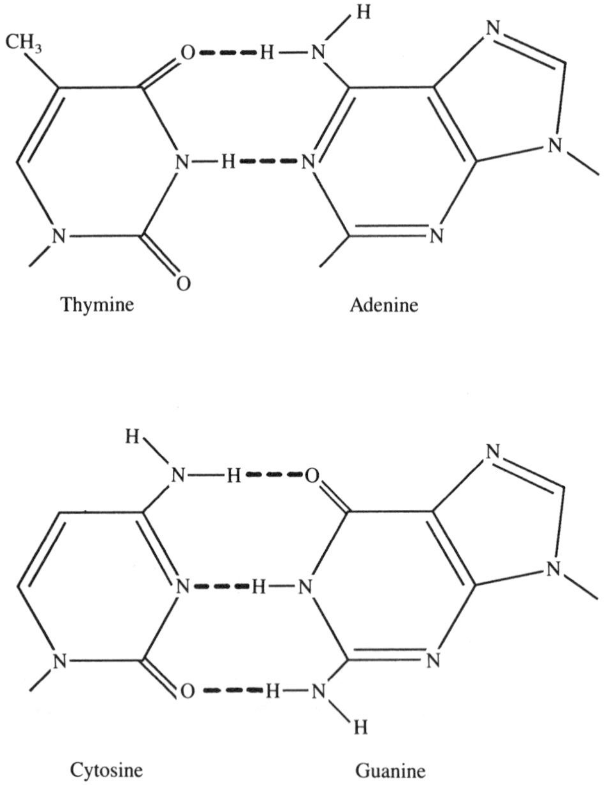

Figure 8-3. Hydrogen bonding between bases as proposed by Watson and Crick. Refer to Figure 6-6 in order to see how the standard diagrams of these molecules must be rotated to generate this model. Note that, as diagrammed here, the outlines of the two pairs are identical; one can be superimposed upon the other. This latter property makes it possible to arrange these base pairs within the DNA molecule in any order (not in the fixed ratio demanded by Levene's model shown in Figure 8-2).

Lukewarm to such data or not, Watson and Crick devised *the* model that accounted for everything then known about DNA: the basic structure of the sugar-phosphate backbone, the two-stranded structure of the molecule, and the one-to-one ratios of adenine to thymine and guanine to cytosine. The model was not arrived at without effort. For example, compare the two purines and two pyrimidines illustrated in Figure 6-6 with Figure 8-3, which shows the chemical bonding between a purine and a pyrimidine base. Note that the outlines of the upper and lower figures in Figure 8-3 can be precisely superimposed one on the other. Figure 8-4 is an exploded diagram

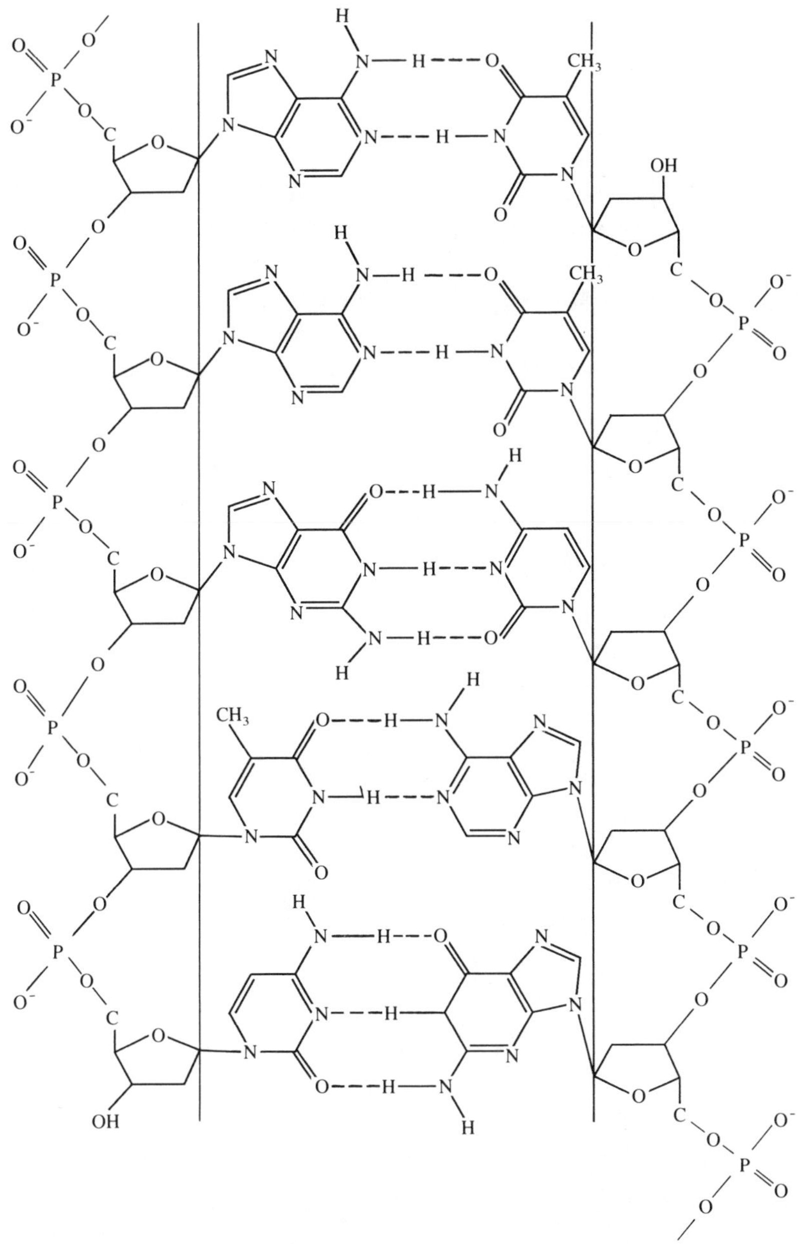

Figure 8-4. An exploded diagrammatic view of a short segment of a DNA molecule. An examination of the sugar-phosphate linkage reveals that the two strands are oriented in opposite (antiparallel) directions. The vertical lines emphasize the identities of the spaces required by all five base pairs, which are (from top to bottom): A-T, A-T, G-C, T-A, and C-G.

of the physical structure of DNA as deciphered by Watson and Crick. The two vertical lines emphasize the *exact* fit of each pair of bases—guanine-cytosine, cytosine-guanine, adenine-thymine, and thymine-adenine—between the sugar-phosphate "backbones." The exact fit reveals that these base pairs can be arranged in any order (unlike the assumption underlying the tetranucleotide model) and can assume any proportion (as Chargaff had suggested) from 100% A-T, 0% C-G to the reverse, including any intermediate value. Furthermore, no matter what the overall composition might be, there is always a one-to-one ratio of adenine to thymine and of guanine to cytosine.

A final point concerning the Watson-Crick model is sketched in diagrammatic form in Figure 8-5 (also see Figure 8-4). Note that the double

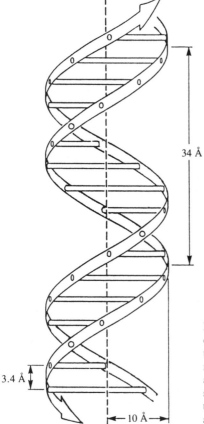

34 Å

3.4 Å

←— 10 Å —→

Figure 8-5. The DNA molecule. Arrows emphasize the antiparallel arrangement of the two strands. The double helix with a major and a minor groove results from forces acting upon the constituent components. The simple rungs that connect the two strands are the base pairs; these pairs are as interchangeable as the rungs in a household ladder.

structure consists of two complementary fibers that run in opposite directions. This illustrates a point that is easily overlooked. From the early days in Morgan's laboratory, *Drosophila* geneticists knew that segments of chromosomes are occasionally rotated 180° relative to their normal sequence. Data on the recombination between gene loci offered the first evidence: gene maps showed reversals of gene order, as illustrated in Figure 8-6. Subsequently, however, the giant chromosomes of larval salivary glands revealed the inverted sequences in exquisite detail. The question posed here is this: Given a need for a physical model such as one that might be used to demonstrate the properties of DNA, how must it be constructed? A colleague of mine tried to use a segmented balsam stick that looked like this:

When he attempted to show his students an inversion, however, he confronted this predicament:

At one position, pegs confronted one another, and two holes were juxtaposed with no pegs to hold them at the neighboring position.

Putting two pegs on one end of each piece and two holes in the other did not help; it merely multiplied the problem by two. My colleague finally arrived at a solution by building his balsam pieces as follows:

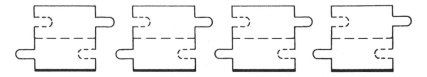

Note the polarity of these pieces! They correspond to the antiparallel strands of DNA. Note, too, that a central piece can be inverted, rotated, and once more inserted between the two flanking pieces. (When it is inverted, the heavy bottom line of the inverted piece comes to lie at the top and the light top line moves to the bottom.) Purely mechanical considerations such as this govern the manipulations that DNA can undergo. Some additional "string puzzles" are illustrated in Figures 8-7 and 8-8.

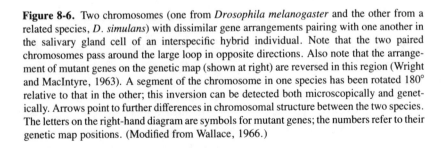

Figure 8-6. Two chromosomes (one from *Drosophila melanogaster* and the other from a related species, *D. simulans*) with dissimilar gene arrangements pairing with one another in the salivary gland cell of an interspecific hybrid individual. Note that the two paired chromosomes pass around the large loop in opposite directions. Also note that the arrangement of mutant genes on the genetic map (shown at right) are reversed in this region (Wright and MacIntyre, 1963). A segment of the chromosome in one species has been rotated 180° relative to that in the other; this inversion can be detected both microscopically and genetically. Arrows point to further differences in chromosomal structure between the two species. The letters on the right-hand diagram are symbols for mutant genes; the numbers refer to their genetic map positions. (Modified from Wallace, 1966.)

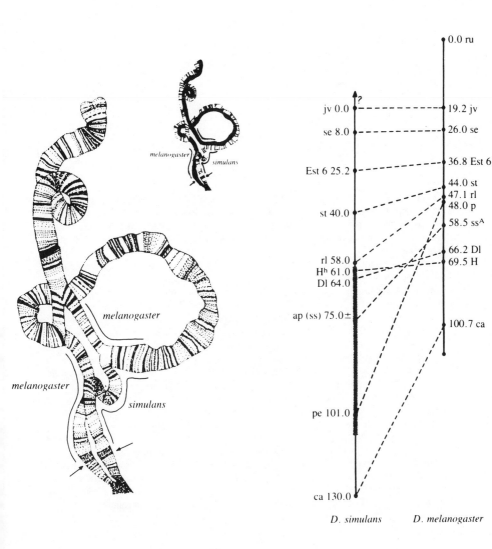

D. simulans   D. melanogaster

Figure 8-7. Certain characteristics of DNA arise as a direct consequence of its structure. In the accompanying diagrams, recombination of DNA segments is assumed to be restricted to regions having identical base sequences. Although these are the customary sites for recombination events, enzymes do exist which permit recombination between nonidentical segments.

I. Recombination between two reverse-oriented repeat segments results in the inversion of the intervening segment. (1) The initial DNA molecule (note the orientation, *A-XYZ-B*). (2) A hairpin loop that results from the pairing of segments *abc* and *cba*. (3) Following an exchange in the repeated segment, the DNA molecule has the orientation *A-ZYX-B*.

II. Recombination between two tandemly oriented repeat segments results in the loss of a circular piece of DNA. 1, 2, and 3 are comparable to those in I except for the outcome of recombination. (4) A circular piece of DNA that contains a segment homologous with one in a second piece may "pair" and, through recombination, be inserted into the second molecule (5). Viral DNA and DNA transcribed from RNA viruses (such as the HIV virus that causes AIDS) become incorporated into the hosts' nuclear DNA in this way.

III. A hint at the complexities that arise from segments of DNA containing more or less identical adjacent segments in reverse order ("palindromic" DNA). (1) Such palindromic DNA can pair as if the two palindromes are in the reverse orientation, as in I(2). (2) Alternatively, the two palindromes may pair as if they are direct repeats of one another, as in II(2). It is left to the reader to determine the consequences of exchanges that might occur at each of the four double-headed arrows. (Note that the two palindromic segments differ [○ and ●] in the region between the letters *c*.) The reader may also convince himself or herself that upon the reinsertion of a circle containing a palindromic sequence, either of two orientations is possible. These alternative orientations may have important consequences with respect to the control of gene action.

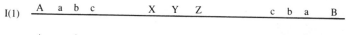

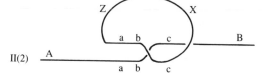

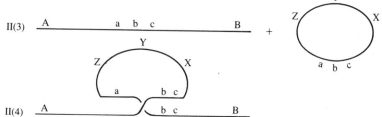

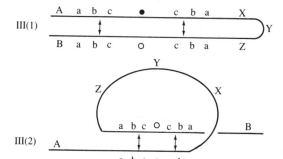

Figure 8-8. A protozoan constructs a functional gene from nine fragments that were initially isolated and arbitrarily assigned numbers 1–9. (a) The nine fragments that possess identical, overlapping terminal segments. (b) A step sequence showing that, except for the two terminal ones, each fragment includes segments of DNA that are identical to those of two others. (c) The final DNA molecule generated by recombination of the overlapping segments. Additional base pairs other than the matching terminal ones are present in the numbered area set off by parentheses; each of these sequences, ranging in length from 15 to 476 base pairs, is unique. This complex procedure for forming a gene accompanies the possession of two nuclei (a characteristic exhibited by protozoans): a micronucleus that is involved in sexual reproduction and a macronucleus that governs metabolic activities. (Modified from Greslin et al., 1989.)

```
GGAGTCGTCAAG  (1) AATC
CCTCAGCAGTTC  (1) TTAG

        AATC  (2) CTCCCAAGTCCAT
        TTAG  (2) GAGGGTTCAGGTA

   GCCAGCCCC  (3) CAAAACTCTA
   CGGTCGGGG  (3) GTTTTGAGAT

CTCCCAAGTCCAT  (4) GCCAGCCCC
GAGGGTTCAGGTA  (4) CGGTCGGGG

  CAAAACTCTA  (5) CTTTGGGTTGA
  GTTTTGAGAT  (5) GAAACCCAACT

 AGGTTGAATGA  (6) 3'TAS
 TCCAACTTACT  (6) 5'TAS

CTTGACGACTCC  (7) ATGTGTAGTAAG
GAACTGCTGAGG  (7) TACACATCATTC

       5'TAS  (8) CTTACTACACAT
       3'TAS  (8) GAATGATGTGTA

 CTTTGGGTTGA  (9) AGGTTGAATGA
 GAAACCCAACT  (9) TCCAACTTACT
```

(a)

5'TAS (8) CTTACTACACAT
3'TAS (8) GAATGATGTGTA

CTTGACGACTCC (7) ATGTGTAGTAAG
GAACTGCTGAGG (7) TACACATCATTC

GGAGTCGTCAAG (1) AATC
CCTCAGCAGTTC (1) TTAG

AATC (2) CTCCCAAGTCCAT
TTAG (2) GAGGGTTCAGGTA

CTCCCAAGTCCAT (4) GCCAGCCCC
GAGGGTTCAGGTA (4) CGGTCGGGG

GCCAGCCCC (3) CAAAACTCTA
CGGTCGGGG (3) GTTTTGAGAT

CAAAACTCTA (5) CTTTGGGTTGA
GTTTTGAGAT (5) GAAACCCAACT

CTTTGGGTTGA (9) AGGTTGAATGA
GAAACCCAACT (9) TCCAACTTACT

AGGTTGAATGA (6) 3'TAS
TCCAACTTACT (6) 5'TAS

(b)

5'TAS (8) CTTACTACACAT (7) GGAGTCGTCAAG (1) AATC (2) CTCCCAAGTCCAT (4) GCCAGCCCC (3) CAAAACTCTA (5) CTTTGGGTTGA (9) AGGTTGAATGA (6) 3'TAS
3'TAS (8) GAATGATGTGTA (7) CCTCAGCAGTTC (1) TTAG (2) GAGGGTTCAGGTA (4) CGGTCGGGG (3) GTTTTGAGAT (5) GAAACCCAACT (9) TCCAACTTACT (6) 5'TAS

(c)

The duplication of DNA

"It has not escaped our notice that the specific [nucleotide base] pairing we have postulated immediately suggests a possible copying mechanism of the genetic material." Those are the words Watson and Crick (1953a, p. 737) wrote in the first report of their DNA model. Later in the same year, at Cold Spring Harbor, the two continued, "We should like to propose . . . that the specificity of DNA self replication is accomplished without recourse to specific protein synthesis and that each of our complementary DNA chains serves as a template or mould for the formation onto itself of a new companion chain" (1953b, p. 128). The significance of these remarks is illustrated in Figure 8-9, which shows how the Watson-Crick model of DNA leads automatically to an understanding of its accurate reduplication during cell division.

At least 20 proteins, including DNA polymerase, are needed for the replication of DNA. Energy is needed as well. These, however, are matters primarily of concern to biochemists and molecular biologists whose interests lie in the study of gene action. The fascinating aspect of DNA replication in our search for the gene is emphasized by repeating Muller's (1961, p. 6) remarks:

> We do not know of any enzymes or other chemically defined organic substances having specifically acting autocatalytic properties such as to enable them to construct replicas of themselves. Neither was there a general principle known that would result in pattern-copying. . . . However, the determiners or genes themselves must conduct, or at least guide, their own replication, so as to lead to the formation of more genes just like themselves, in such wise that *even their own mutations become incorporated in the replicas* [emphasis added]. . . . That is, this genetic material must underlie all evolution based on mutation and selective multiplication.

Furthermore, "Natural selection based on the differential multiplication of variant types cannot exist before there is material capable of replicating itself and its own variations, that is, before the origination of specifically genetic material or gene-material" (p. 7).

Muller's comments are worth reading more than once. They show how the postulated requirements of genetic material are met by properties inherent in the structure of DNA: the ability to replicate, the ability to change, and the ability to replicate the mutated form. Furthermore, once

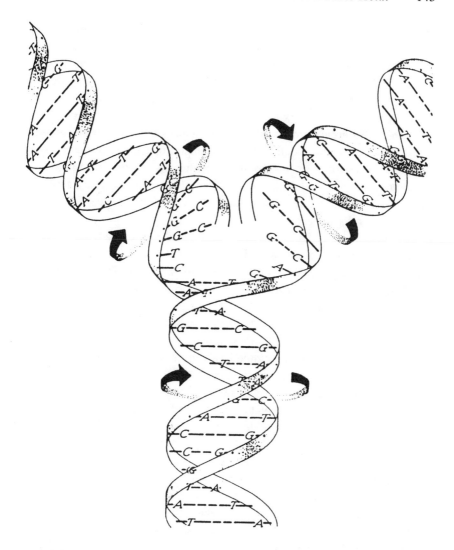

Figure 8-9. Replication of DNA. The problems associated with the uncoiling of the parental double helix and the coiling of the daughter strands are not trivial; at least 20 proteins are required for replication. Two of these form *gyrase*, the enzyme responsible for uncoiling the double helix.

the implication of these comments is grasped, it becomes obvious why many urge extreme caution in the artificial manipulation or "engineering" of DNA. The release of altered DNA into our environment in no way corresponds to the release of any other chemical. DNA can be replicated, it can change, and the changed form can be replicated. In Muller's words, DNA "must underlie all evolution based on mutation and selective multiplication." He argued under other circumstances that properties defining *life* correspond to those defining DNA (genetic material); all other physiological, behavioral, or psychological definitions of life refer to characteristics now associated with life, to be sure, but they are characteristics useful in protecting and perpetuating DNA, and hence were acquired through natural selection.

The replication of DNA, the act Watson and Crick asserted "has not escaped our notice," is illustrated in Figure 8-9. Here the two strands of DNA have been separated, and on each of the single strands a newly synthesized strand is being formed, thus restoring the double helix of two-stranded DNA. Casual examination reveals that the two new strands are identical and that each new double helix is identical to the double strand that existed earlier. Each old strand acts as a template that specifies the sequence of purine and pyrimidine bases in the one currently under construction. Like does not specify like; complementation, not identity, governs the matching bases. The two strands of double-stranded DNA are not identical; they are complementary. The old and new strands of replicating DNA molecules form two double helixes identical (barring errors) to the original molecule.

By what mechanism does a double helix become untwisted so that the two strands are isolated from one another? And how does each newly replicated strand form a double helix once more? You can ask; those are good questions. But they are also questions best addressed to a specialist in molecular mechanics. For illustrative purposes Crick frequently referred to the twists found in hempen ropes. Personally, I am not into ropes and their manufacture, but to visualize many of the problems associated with DNA and its coiling (or uncoiling), a length of half-inch hempen rope (reduced to two strands rather than the more normal three) can be useful.

One might still ask, nevertheless, about the process of replication. The diagram shown in Figure 8-9 was drawn for illustrative purposes only. One might imagine that the two newly created complementary strands may pair in some manner so that replication leads to the conservation of the old two-stranded DNA and a spanking new one consisting of two newly created single strands. Or, alternatively, the new strands being formed might jump back and forth between the old ones so that, with many fractures, new and old segments form a linear mosaic. In that case, there would be no wholly new or wholly old strand.

The above question was attacked and answered by Matthew S. Meselson and Franklin W. Stahl using a heavy isotope of nitrogen, ^{15}N. Bacteria (*E. coli*) were grown in media containing heavy nitrogen which was incorporated in considerable quantities into the nitrogenous bases (purines and pyrimidines—A, T, G, and C) of the bacterial DNA, the bacterial chromosome. Once a culture of cells whose chromosomes were thoroughly labeled

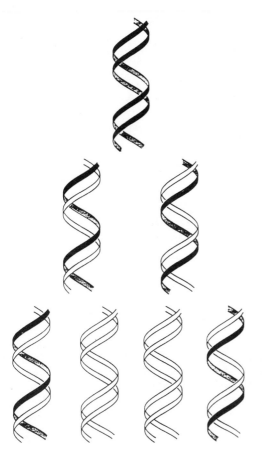

Figure 8-10. The distribution of parental atoms among daughter DNA strands. The original parental strands are shown in black. Each goes intact to one of the daughter strands. During the next round of replication, each goes intact to one of two of the now-four DNA strands. The remaining two of the four strands were copied from the "white" strands of the first-generation daughter molecules and so are entirely white.

with ^{15}N was established, an excess of ^{14}N-containing nutrients was added to the medium. DNA made after this time would consist almost exclusively of purines and pyrimidines containing ^{14}N, light nitrogen. By extracting the DNA from bacteria sampled at various times over four generations (about one and a half hours!) and subjecting it to ultra (i.e., extremely high speed) centrifugation, the relative amounts of DNA of different densities

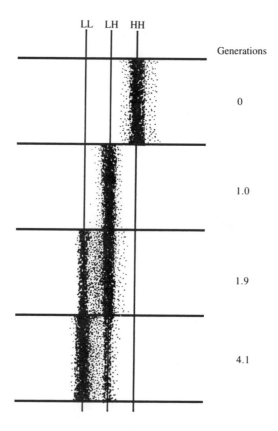

Figure 8-11. Experimental observations made by Meselson and Stahl (1958) using ^{15}N-labeled substrates, ultracentrifugation of cesium chloride solutions (to separate the LL, LH, and HH double-stranded DNA as explained in Box 8-1 and the text), and photographs based on UV absorption by DNA (see Figure 6-8). Ruled lines have been superimposed on the diagram so that observations on LL, LH, and HH bands can be more easily compared with the interpretation presented in Figure 8-10. Note that in the fourth generation, of 16 double helices, 14 should be LL and 2 LH; this 7:1 ratio is revealed in Meselson and Stahl's observations. (L = light [^{14}N-labeled] DNA strands, and H = heavy [^{15}N-labeled] strands.)

were determined. Figure 8-10 illustrates Meselson and Stahl's conclusions: The strands of the old DNA molecule (heavy with ^{15}N) separated and each specified the construction of a new ^{14}N-labeled strand. The new molecule—half heavy nitrogen and half light nitrogen—had an intermediate density. In subsequent generations, ^{14}N-containing DNA replicated by making new ^{14}N-containing strands, and the total amount of light DNA accumulated; a smaller and smaller proportion of all DNA still consisted of "hybrid" molecules of intermediate density. The observations on which these conclusions were reached are illustrated in Figure 8-11. I leave it to the reader to interpret Figure 8-11 and arrive at the conclusion illustrated in Figure 8-10. Especially important features are manifest at 0, 1.0, 1.9, 3.0, and 4.1 generations, as illustrated.

Box 8-1. The ultracentrifuge

The centrifuges that resembled inverted eggbeaters and that were clamped to laboratory tables or windowsills in many of my undergraduate laboratories are collectors' items today. I have not seen one for years. Nevertheless, they were useful devices that allowed one to spin down some crude preparation, pour off the supernatant, replace it with another liquid (distilled water, Ringer's solution, or alcohol—whatever the protocol demanded), resuspend the material, spin it down once more, and repeat until the original liquid was thoroughly displaced.

The term *spin down* refers to the tendency of heavy materials to settle out in a liquid suspension. Sand settles out very quickly when shaken and suspended in water; the white flakes in the snow scenes enclosed in plastic globes found in novelty stores settle out much more slowly. The sand and the artificial snowflakes settle because of gravity; because of centrifugal force, blood rushes to the head of an aviator doing outside loops (the outcome is known as "red out").

Elementary physics courses are built around a simple equation: $F = ma$. Force equals mass times acceleration. A body that travels in a circular path like a rock on a biblical-style sling exhibits acceleration simply by not flying off in a straight line. In this case, $a = v^2/r$, where v is the velocity of the rock and r is the length of the restraining thong.

Given that all else is the same between two centrifuges, the force that acts on objects in spinning tubes—let's say, on heavy objects that tend to settle out—is proportional to the square of the revolutions per minute. The old hand-operated centrifuge may have turned 1000 times per minute. Meselson and Stahl used an ultracentrifuge that spun at a rate of 44,770 revolutions per minute. The force of the ultracentrifuge on suspended particles, then, was nearly 2000 times greater than that of the hand-operated kind, provided that the diameters (or radii) of the circle in which objects spun were comparable. Under such tremendous forces, small differences between the densities of suspended material and the suspending liquid cause these materials to spin down or float in a relatively short time.

Meselson and Stahl used a second trick that in 1958 called for explanation even in a professional journal. They suspended their DNA preparations (which contained ^{14}N and ^{15}N isotopes of nitrogen) in a cesium chloride (CsCl) solution. This salt solution is particularly useful for many purposes because it tends to "pack," thus creating a solution that ranges from high density at the bottom of the centrifuge tube to a lighter density at the top surface. Consequently, the DNAs of different density, because of the ^{15}N and ^{14}N they contain, do not merely settle out at different rates (like sand grains and bread crumbs in a glass of water); they come to rest and form bands when they reach the point in the CsCl gradient that matches their own density (i.e., where centrifugal and buoyant forces are equal). They spin down so far, and no farther. DNA containing ^{15}N in both of its two strands (HH, where H means "heavy") forms one band; DNA containing ^{15}N in one strand and ^{14}N in the other (HL, where L means "light") forms another band not quite as far down the centrifuge tube as the first; and DNA both of whose strands have been built of ^{14}N-containing purines and pyrimidines (LL) is the lightest of all. These are the three bands shown in Figure 8-11.

9 The Genetic Code and Related Matters

At times one writer's concluding remarks are appropriate for another's introductory ones. Consequently, Francis Crick's (1966, p. 9) conclusion to his introductory comments serves me well in introducing this chapter: "One can say . . . that the elucidation of the genetic code . . . is, in a sense, the key to molecular biology because it shows how the two great polymer languages, the nucleic acid language and the protein language, are linked together."

Language is perhaps a novel term for one acquainted with rather elementary biology. Communication has only recently been recognized as an essential part of the biological sciences. A colleague who during the 1950s determined experimentally how a mother wasp arranged her individual brood chambers so that her progeny, as they emerged from their pupal cases, would not interfere with one another, had the title of his report ("Communication between Two Generations of Wasps") rejected by an editor. Perhaps the acceptance of communication as a real phenomenon stems from Karl von Frisch's (1950) analysis of bee dances, a topic that lies far from our search for the gene.

At this point, several factual points have been established:

1. DNA is the genetic material; the gene is DNA.
2. DNA is a long, fibrous molecule. As Frank Stahl (1964, p. 26) wrote in an introductory text, "The DNA molecule is 20 Å wide and terribly long." The chromosome of *E. coli* is 1 mm long; the four chromosomes of *D. melanogaster* contain 16 mm of DNA.

3. The gene is responsible for protein synthesis. This is implicit in the one gene–one enzyme hypothesis. It is also implicit in the single amino acid substitution that differentiates sickle-cell hemoglobin from normal hemoglobin.

4. Protein molecules, despite the folded, globular, or sheetlike configurations they eventually assume, have a long, fibrous primary structure in which one amino acid is attached to another in single-file order.

How does a long molecule of DNA specify the (also long) sequence of amino acids in a protein molecule? By direct contact was one early guess, but no workable scheme was forthcoming. Those studying the biochemistry of gene action learned that the synthesis of a new protein was inhibited if RNA synthesis was also inhibited. The needed RNA (the messenger RNA, or mRNA) is transcribed from a localized portion of DNA and then, through the operation of one of the cell's most complicated instruments—the ribosome—specifies point by point what amino acid will be attached next to the ever-lengthening polypeptide chain.

I am tempted to cover all facets of protein synthesis, but these are chemical or biophysical events related to how genes carry out their functions, and thus outside the realm of our search. A more pertinent question is: "How can one show that a molecule of RNA is transcribed from a particular segment of DNA? In a series of experiments, Sol Spiegelman and his colleagues (see, for example, S. A. Yanofsky and Spiegelman, 1963) demonstrated that two species of RNA molecules do not compete in attaching themselves (annealing) to complementary DNA. Thus, when RNA 1 has saturated DNA so that no more can anneal, RNA 2 anneals at the same rate as if RNA 1 were absent. The reverse is also true: saturation of DNA with attached RNA 2 does not affect the rate at which RNA 1 anneals. Thus, "no evidence of competitive interaction between the two for the same sites can be detected" (Yanofsky and Spiegelman, 1963, p. 543). The DNA-RNA complexes involve two physically separated regions of the DNA.

The first clear advance in understanding how DNA (via mRNA) specifies amino acid sequences came through the construction of an artificial mRNA that contained only uracil (poly U, or UUUUUU . . .). When poly U was added to extracts made of crushed *E. coli,* a polypeptide chain containing only phenylalanine was constructed: phe-phe-phe-phe- . . . , polyphenylalanine. The *E. coli* ribosomes attached themselves to the poly U artificial mRNA and proceeded, with the aid of the enzymes and energy sources present in the extract, to string together a series of phenylalanines (Nirenberg and Matthaei, 1961).

In rapid sequence, other investigators learned that poly A (artificial mRNA containing only adenine) led to the synthesis of a polypeptide consisting only of lysine (polylysine), and poly C led to the synthesis of polyproline. More complex poly chains were made by chemical manipulations and tested.

Artificially synthesized poly AC containing 47% A and 53% C would be expected to contain the following proportions of *pairs* of bases: AA, 22.1%; AC, 24.9%; CA, 24.9%; and CC, 28.1%. Triplets of the two bases A and C would be expected as follows: AAA, 10.4%; AAC, 11.7%; ACA, 11.7%; CAA, 11.7%; CCA, 13.2%; ACC, 13.2%; CAC, 13.2%, and CCC, 14.9%.

The polypeptide chain that resulted from the use of poly AC as a messenger RNA had the following composition: lysine, 10.8%; asparagine, 11.6%; glutamine, 9.3%; threonine, 26.3%; histidine, 9.4%; and proline, 32.6%. Clearly, the proportions of amino acids more nearly coincide with the proportions of triplets; when the proportions deviate, as in the cases of threonine and proline, the discrepancy seems to involve a factor of two, or even three.

Experiments such as these led to the final solution to the genetic code—that is, to the revelation of which combinations of U, A, C, and G specify each of the 20 amino acids used in protein synthesis. With only 4 purine and pyrimidine bases, it had been clear from the outset that 1 base could not specify 1 amino acid; there are too many amino acids. Nor could 2 bases suffice, because only 16 combinations, not 20, could be formed. Overlapping combinations in which a single change in DNA would, of necessity, affect several amino acids were ruled out. Hence, triplets were favored; they can form $4 \times 4 \times 4$, or 64, combinations.

Figure 9-1 presents the genetic code in the form of a dictionary that states which triplet specifies which amino acid. Many combinations can specify the same amino acid: there are six combinations for leucine, arginine, and serine; four for many others; three for isoleucine; two for phenylalanine, tyrosine, cysteine, glutamine, and others; and one for tryptophane and methionine. Three triplets are signals that call for the termination of the growing polypeptide chain: UAA, UAG, and UGA. The codon for methionine also seems to be a start signal; many proteins have methionine as the

	U	C	A	G	
U	UUU ⎫ Phe UUC ⎭ UUA ⎫ Leu UUG ⎭	UCU ⎫ UCC ⎪ Ser UCA ⎪ UCG ⎭	UAU ⎫ Tyr UAC ⎭ UAA Stop UAG Stop	UGU ⎫ Cys UGC ⎭ UGA Stop UGG Trp	U C A G
C	CUU ⎫ CUC ⎪ Leu CUA ⎪ CUG ⎭	CCU ⎫ CCC ⎪ Pro CCA ⎪ CCG ⎭	CAU ⎫ His CAC ⎭ CAA ⎫ Gln CAG ⎭	CGU ⎫ CGC ⎪ Arg CGA ⎪ CGG ⎭	U C A G
A	AUU ⎫ AUC ⎬ Ile AUA ⎭ AUG Met	ACU ⎫ ACC ⎪ Thr ACA ⎪ ACG ⎭	AAU ⎫ Asn AAC ⎭ AAA ⎫ Lys AAG ⎭	AGU ⎫ Ser AGC ⎭ AGA ⎫ Arg AGG ⎭	U C A G
G	GUU ⎫ GUC ⎪ Val GUA ⎪ GUG ⎭	GCU ⎫ GCC ⎪ Ala GCA ⎪ GCG ⎭	GAU ⎫ Asp GAC ⎭ GAA ⎫ Glu GAG ⎭	GGU ⎫ GGC ⎪ Gly GGA ⎪ GGG ⎭	U C A G

Figure 9-1. The genetic code. Sixty-one of the 64 possible triplets of bases correspond to one or another of the 20 amino acids. The remaining three—UAA, UAG, and UGA—are signals that terminate the synthesis of the growing polypeptide chain. Transfer RNA (tRNA) molecules that correspond to these three stop codons do not exist. A tRNA molecule that normally reads UAC may, through a rare mutation, come to read UAA; in that case, a tyrosine molecule would be inserted in the growing chain at the site where the chain should have ended. This point is made here to emphasize once more the variety of events that may alter the normal course of protein synthesis. Borrowing a phrase from Francis Crick (1966), this table is the dictionary—or Rosetta stone—that links "the two great polymer languages, the nucleic acid language and the protein language."

first amino acid (in others, the methionine is removed *subsequent* to protein synthesis).

Some of the excitement scientists working to decipher the code felt, and the extent to which various workers cooperated in their common effort, can be sensed in F. H. C. Crick's (1966, p. 3) reminiscences:

> This is an historic occasion. There have been many meetings about the genetic code during the past ten or twelve years but this is the first important one to be held since the code became known. When I came to the States early in 1965 I brought with me tentative allocations for many of the 64 triplets, based mainly on the early work of Leder and Nirenberg, the results from the random polymers and the mutagenesis data. I telephoned Marshall Nirenberg, who told me of his latest allocations. A little later I saw Gobind Khorana and heard the first results he was getting using polymers with repeating sequences. I also visited George Streisinger and was told about the preliminary amino acid sequences due to a phase shift in the phage lysozyme. From all this we were able to work out the meaning of several of the remaining doubtful triplets. By March 1965 the great majority of triplets had been unambiguously identified and just a few remained unallocated. It was a most exciting occasion for me, travelling about the country and seeing how the various lines of evidence fitted together.

Twenty years later, because of the commercial value attached to many of the pharmaceutical and agricultural products made possible by research in molecular biology, preliminary results are sometimes no longer shared over the phone or during casual visits to others' laboratories. Molecular research is often technological research, and today's information is frequently labeled "proprietary."

Mutation

Twice—when he referred to Leder and Nirenberg and then to Streisinger—Crick touched on the matters of mutagenesis and mutation with respect to the aid they provided in breaking the genetic code. Because we are emphasizing that DNA is something to be read, a simple sentence in English may serve to illustrate mutations.

Consider the following sentence:

The mad man cut his arm.

The substitutions of individual letters for one another could alter this sentence in several ways:

a → o (1)

The mod man cut his arm.

a → o (2)

The mad mon cut his arm.

a → o (3)

The mad man cut his orm.

m → s (1)

The sad man cut his arm.

m → s (2)

The mad san cut his arm.

m → s (3)

The mad man cut his ars.

Obviously, some alterations result in a sentence that remains intelligible but has an altered meaning (a *sad* man is not the same as a *mad* man). Others, however, generate nonexistent words (mon and san). In the genetic code, these simple substitutions of one base for another can arise spontaneously because purines and pyrimidines on rare occasions take on unusual chemical structures (tautomeric shifts) by chance alone, thus allowing adenine to pair with cytosine (A-C rather than the anticipated A-T) and guanine to pair with thymine (G-T rather than G-C). When DNA that contains such errors replicates, the two daughter strands differ at the site of the mismatched pair:

Error-containing parental strand	Daughter strands	
A-C	A-T	G-C
G-T	G-C	A-T

Clearly, both daughter strands cannot be correct; one will lead to an error in protein synthesis. Recalling the single amino acid substitution that altered normal hemoglobin (containing glutamic acid) to sickle-cell hemoglobin (containing valine), one can see from Figure 9-1 that a simple "typographical" error could change GUA to GAA or GUG to GAG; either change would result in the observed substitution. And, just as a mad man and a sad man may differ considerably, so do the two types of hemoglobin.

Crick's reference to Streisinger mentions a different sort of mutation, "frameshift." Once more, consider the sentence

The mad man cut his arm.

The spaces between the words allow us to see the words more clearly; otherwise we might write

Themadmancuthisarm.

Let's now delete the first *e,* and once more print the sentence in three-letter segments:

Thm adm anc uth isa rm

Here is a sentence that has suffered a frameshift mutation. In English, it now has no meaning. In Streisinger's experiments, the frameshift mutation gave rise to a new series of amino acids, thus enabling him to infer the DNA code words.

A second frameshift mutation can be imposed on the nonsensical sentence printed above by removing the leftmost *a.* The result is

Thm dma ncu thi sar m.

Still nonsense, but now, if we remove the second *m:*

Thm dan cut his arm

Suddenly, familiar words reappear. This is comparable to one of the crucial experiments that demonstrated that the genetic code is really (not just theoretically) based on triplets of bases: If three bases are deleted at the left (not necessarily adjacent to each other) or three bases added at the left, then, just to the right of the third alteration, the polypeptide chain once more possesses the same amino acids that it had originally. The reading frame covers three bases at a time (as our eye sought out three-lettered words such as *the, man,* and so forth); three single base deletions or three single base additions put the reading frame back on track once more. Figure 9-2 illustrates a variety of mutations using biological rather than alphabetical symbols.

Colinearity

My discussion of DNA code words and of amino acids in polypeptide chains has assumed that the sequence of one corresponds directly to the

DNA	mRNA	Polypeptide

Nonmutated:

```
ATTCTGACTTGC
TAAGACTGAACG        AUUCUGACUUGC        Ile-Leu-Thr-Cys
```

Missense mutation; synonymous:

```
           *
ATTTTGACTTGC
TAAAACTGAACG        AUUUUGACUUGC        Ile-Leu-Thr-Cys
    *
```

Missense mutation; nonsynonymous:

```
      *
ATTCTGGCTTGC
TAAGACCGAACG        AUUCUGGCUUGC        Ile-Leu-Ala-Cys
      *
```

Mutation to "stop" codon:

```
          *
ATTCTGACTTGA
TAAGACTGAACT        AUUCUGACUUGA        Ile-Leu-Thr* (terminate)
          *
```

Frameshift mutation (caused by insertion of new base pair at left):

```
*
GATTCTGACTTGC
CTAAGACTGAACG       GAUUCUGACUUGC       Asp-Ser-Asp-Leu
*
```

Figure 9-2. Some mutational events and their consequences relative to polypeptide synthesis. The messenger RNA is transcribed in each case from the lower of the two DNA strands. Note that the first missense mutation (synonymous) leaves the polypeptide chain unchanged; the second one (nonsynonymous) does not. Asterisks indicate the site of the mutational changes.

sequence of the other. This assumption is by no means confirmed by anything we have learned so far. One can imagine, for example, that a complexly folded messenger RNA is related to a complexly folded protein such that an altered RNA code word at position 5 may, in fact, alter the 23rd amino acid in the corresponding polypeptide chain. This possibility may seem unreasonable, but, to argue in its favor, consider that in holding oxygen within a molecule of hemoglobin, the important amino acids (the *active site*) may be scattered from one another at considerable distances; the folding of the protein brings these amino acids together in juxtaposition so that they carry out their proper function.

The demonstration of colinearity of DNA and amino acid sequences that I am about to present requires, first, homage to large numbers—*really* large numbers! Mendel counted hundreds or low thousands of peas in establishing his classical conclusions. Morgan's group raised and counted tens of thousands of flies during its entire tenure at Columbia University. Geneticists who switched to bacteria and phage switched as well to procedures that led to the detection of extremely rare events. Billions, even tens of billions, of bacteria can be spread upon an ordinary petri dish (about 4 inches in diameter). The presence of streptomycin or another antibiotic in the nutrient medium will reveal the 1, 2, or 3 resistant bacteria among the billions that are not; all other bacteria die. This is one of many "selective" techniques that reveal the needles in extremely large haystacks. Indicator media provide another selective technique; rare individual bacteria with abnormal (or at least different) metabolisms can be identified by indicator media that change color beneath the appropriate bacterial colonies. A mutation that would occur only once every two generations in the population of the United States (once in a billion gametes) would appear several times on a small petri dish that was spread initially with 10^{10} bacteria.

Charles Yanofsky and his colleagues (Yanofsky et al., 1964) took advantage of these considerations in creating a genetic map of mutations that destroyed the function of a particular protein in *E. coli*. Because these mutations were all within the region of DNA that specified this protein, the map distances were exceedingly small. Nevertheless, by studying large numbers of tested bacteria, reliable distances were obtained for 16 mutants.

Next came the task of identifying the amino acid that was altered in each of the 16 mutant proteins. Earlier, I pointed out that sickle-cell hemoglobin differs from normal hemoglobin by an amino acid substitution in the sixth site (residue). That analysis can, of course, be extended so that the amino acid present at any site in any protein can be determined. Having analyzed the linkage relationships by genetic means and the position of the changes in the protein molecule itself, Yanofsky and his colleagues showed that the closer two changes are genetically, the nearer the two amino acid substitutions are in the protein molecule; similarly, the more remote the changes are according to the genetic test, the farther removed from one another are the physical changes in the molecule. "It would appear," the authors concluded, ". . . that distances on the genetic map are representative of distances between amino acid residues in the corresponding protein" (p. 271). The two macromolecules are colinear. Indeed, the experimental data suggest that each base pair lengthens the map distance by 0.013 map units.

The gene: A new look

Sheer numbers have given us a view of the gene that we did not have before—when we started this search, for example. They have provided us with a microscope with which we can see the gene at close range. Details that earlier were invisible now loom before us. Gulliver, traveling in the land of Brobdingnag, whose inhabitants were gigantic (for them 12 inches was the equivalent of 1 inch for a human being), told of his tabletop view of a wet nurse caring for her charge:

> I must confess no object ever disgusted me so much as the sight of her monstrous breast, which I cannot tell what to compare with, so as to give the curious reader an idea of its bulk, shape and colour. It stood prominent six foot, and could not be less than sixteen in circumference. The nipple was about half the bigness of my head, and the hue both of that and the dug so varified with spots, pimples and freckles, that nothing could appear more nauseous: for I had a near sight of her, she sitting down the more conveniently to give suck, and I standing on the table. This made me reflect upon the fair skins of our English ladies, who appear so beautiful to us, only because they are of our own size, and their defects not to be seen but through a magnifying glass, where we find by experiment that the smoothest and whitest skins look rough and coarse, and ill coloured.

Seymour Benzer, an early member of the Phage Group, took the first close modern look at the gene. The term *gene* had already acquired several meanings. First, for example, it had been an element that exhibited Mendelian inheritance. Mendel himself had followed the inheritance of genes (Merkmalen) for pea color, pea shape, plant height, flower position, and other aspects of *Pisum sativum*. Each gene that he chose for his studies was represented by two alleles (tall versus short, yellow versus green, and round versus angular, to name three). In effect, Mendel's gene was a unit of physiological function.

Linkage presents a complication. Chromosomes are far less numerous than factors that exhibit Mendelian ratios; consequently, many genes must reside on a single chromosome. Genes that are far removed from one another recombine rather freely; hence, it is not difficult to prepare a genetic map, as Sturtevant did, showing the location of the various genes. But as distances become small, identifying genes becomes more difficult. If linkage is absolute, one can speak only of one gene, even though it may have two seemingly unrelated phenotypic effects such as abnormal wing shape and sterility. On the other hand, what is meant by "absolute linkage"? What seems absolutely linked to a person willing to examine 6000 gametes may not prove so to one willing to examine 100,000 gametes. The gene, in this case, has acquired a new definition: it is a unit of recombination.

The discovery that X rays induce genetic change (i.e., are mutagenic) gave rise to still another definition for the gene: the unit of mutation. Consider, for example, a gene with the two effects mentioned above: a mutant gene causes both abnormal wing shape and sterility. Imagine, too, that no one ever obtained a recombinant chromosome in which these two effects were separated. What, then, does one say when exposure of the normal gene to X rays results on rare occasions in mutations that cause abnormal wing shape but leave fertility unchanged or, conversely, cause sterility without altering wing shape? Having shown that these X-ray-induced mutations do in fact occupy the same position on the chromosome as the mutation that causes both abnormalities, one claims that two tightly linked genes are found at that site (those sites)—too close to be separated by recombination but demonstrably separable by mutagenesis.

Thus, when Benzer (1955) started his studies, the gene had already assumed three facets. His ultrafine analysis of a single "gene" in bacteriophage led him to a more precise definition of each of the three facets of the gene than his predecessors had been able to give. In fact, he renamed each facet. The unit of mutation—the *muton*—can be as small as a single base pair substitution (recall the substitution of valine for glutamic acid in hemoglobin) or can involve the loss of many base pairs. The unit of recombination—the *recon*—necessarily involves two adjoining base pairs because the recombination occurs between such adjoining base pairs in the DNA molecule. The unit of function (white versus red eyes in *Drosophila;* tall versus short pea plants)—the *cistron*—involves the stretch of DNA responsible for specifying a functional protein (or RNA molecule). *Cis*-emphasizes that the proper functioning depends upon a single DNA molecule. If two properly constructed proteins are required for a certain purpose, one can be made by one DNA molecule and the other by a second DNA molecule; the two DNA molecules are then said to *complement* one another. Thus, two defective phage particles can successfully infect a bacterium and produce nondefective phage progeny. The gene products of two defective phage particles whose defects lie within the *same* cistron, however, cannot complement one another; consequently, the two—as a pair—cannot, as a rule, successfully infect a bacterium.

Benzer's terminology did not displace the older, more familiar four-letter word *gene*. Why? In part because, even in science, a small amount of ambiguity is useful. Take the notions of "life" and "living." Dogs and carrots live in such obviously different ways that one person has suggested that we say "a dog dogs" and "a carrot carrots" when we normally say "a dog lives" and "a carrot lives." Excess precision does not always facilitate communication; this is not to deny, however, that gross imprecision hinders communication. I believe, too, that matters discussed in the following chapters will reveal that even Benzer's carefully phrased terminology failed to capture all of the nuances surrounding the word *gene*.

Box 9-1. The DNA game

The material covered in Chapter 9 is intriguing to geneticists, but it may be less appealing to others. As a geneticist, even one who is not directly concerned with molecular genetics, I enjoy mulling over the mechanical problems posed by the filamentous molecules of DNA and proteins. Many people, even many biologists, may get little pleasure from doing so. For those people, especially those who have difficulty visualizing chemical molecules and manipulating them mentally, I present the following jigsaw puzzle game to illustrate much of what I discussed in the preceding pages.

The four nitrogenous bases that make up DNA are the purines adenine (A) and guanine (G) and the pyrimidines thymine (T) and cytosine (C). RNA contains the pyrimidine uracil (U), which complements adenine during RNA synthesis or when RNA pairs with itself (or another RNA molecule) to make double-stranded RNA.

In Figure 9-3, the four (five) bases are illustrated floating in space in random orientations. The important configuration, the one postulated by

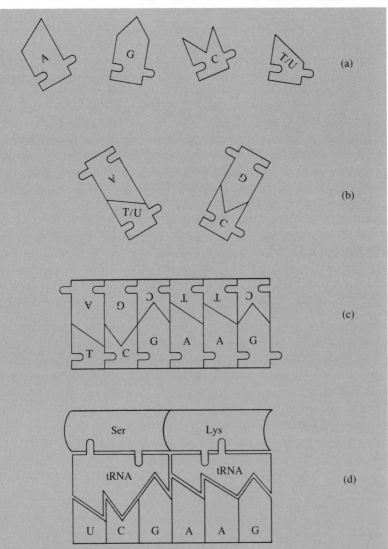

Figure 9-3. The pieces of a DNA jigsaw puzzle game. (a) The five nitrogenous bases (purines and pyrimidines); the same piece serves for T (in DNA) and U (in RNA). (b) The identical shapes of A-T (or T-A) and G-C (or C-G) pairs. (c) Because the outlines of these pairs are identical, they can be arranged in any sequence. (d) The twofold task of tRNA: on the one hand (bottom of the tRNA molecule), it must recognize the three-base codon in mRNA; on the other (top of the tRNA molecule), it must recognize the amino acid specified by the codon. The convex and concave ends of the amino acid pieces emphasize that the polypeptide chain also has a polarity with an NH_3^+ group at one end and a COO^- group at the other (these form the peptide bonds that link the amino acids).

Watson and Crick, is the pair formed by the union of A and T or C and G. Note that these pairs are of the same shape and occupy the same amount of space. Hence, they can be arranged in any sequence, as shown in the center diagram. Any one of the illustrated base pairs could have been drawn in the reverse orientation without disrupting the sequence of interlocking pairs. Had the $\frac{T}{A}$ pair that is second from the right been inserted as $\frac{A}{T}$, the physical dimensions of the DNA molecule would not have been affected. Note that the interlocking pegs and holes must be placed as they are drawn if the pairs are to be rotated and inverted. The outcome is two strands (an upper one shown in the figure as AGCTTC and a lower one shown as TCGAAG) that are antiparallel: the pegs in the two strands point in opposite directions.

Messenger RNA (mRNA) is transcribed using one strand of DNA as a model; RNA in Figure 9-3 was transcribed from the upper strand of DNA, hence it has the same composition as the lower DNA strand except that uracil (U) replaces thymine (T). Consecutive sequences of three bases in the mRNA specify the amino acid that will be inserted into a growing polypeptide chain. A second RNA, transfer RNA (tRNA), is the intermediary in this process. A molecule of tRNA has two tasks: first, to recognize the triplet of bases in the mRNA; second (actually, an earlier task), to recognize and carry the amino acid that corresponds to the mRNA triplet code.

The first task is solved in the puzzle diagram by making the bottom edge of each tRNA molecule irregular so that it complements a set of three bases. On the left, the tRNA complements (or fits, if we speak as puzzle solvers) UCG, the code word for serine; on the right, the tRNA molecule fits AAG, the code word for lysine.

The relationship between the mRNA code word and the proper amino acid depends upon the second property of tRNA: it must have recognized the correct amino acid. This has been solved in the puzzle by placing a peg and a hollow at any two of five specified positions at the top of the tRNA piece. The peg could be at any one of five positions; in each instance, the hollow could be at any one of the four remaining ones. Hence, there are 5×4, or 20 possible combinations of pegs and holes—one for each amino acid. In the diagram, serine has a hollow and a peg that fit those of the corresponding tRNA; lysine, on the other hand, fits the tRNA at the right. Note that polypeptide chains have a polarity with a free amino (NH_3^+) group at one end and a carboxy (COO^-) group at the other; for this reason, each amino acid is shown as convex at one end and concave at the other.

I believe that virtually everything discussed in Chapter 9 can be identified in these puzzle diagrams. Molecules of tRNA that fit UCU, UCC, and UCA all have the top peg at position 1 and the hollow at position 4 because these codons also specify that serine be added to the polypeptide chain. The tRNA molecule that fits the AAA codon has a peg at position 3 and a hollow at position 2 because AAA also codes for lysine. No transfer RNAs would be constructed to fit UAA, UAG, or UGA; these are stop codons that terminate the growth of the polypeptide chains.

Box 9-2. The gene in evolution

On several occasions I have cited early geneticists' claim that the physical basis of the gene must be a chemical that is able (1) to reproduce itself (replicate), (2) to change (mutate), and (3) to reproduce the changed form as faithfully as the original one. In this chapter mutation has been examined in some, although not extensive, detail. In conclusion, I want to say something about gene changes over extended periods of time.

First, I consider a "useful" mutation, the mutation by which a bacterium (*E. coli,* for example) acquires resistance to the antibiotic streptomycin. If several billion bacteria are spread over a streptomycin-containing medium in a petri dish, most of them are killed by the antibiotic. Here and there, perhaps two to five per plate, a colony of bacteria will survive. If these bacteria are respread on streptomycin-containing medium, they all grow; the petri dish becomes covered with a carpet of bacteria.

If many independently arising strains of streptomycin-resistant bacteria are collected and spread on a plain medium (containing no streptomycin), a considerable proportion of these resistant bacteria are *unable* to grow. These strains are both streptomycin resistant and streptomycin *dependent*. Such strains might thrive in an environment laced with streptomycin (as it may well be in countries that allow the sale of streptomycin over the drug store counter without a doctor's prescription) but not in one free of streptomycin.

Next, I add a related example based on *E. coli*. In a wild-type strain that can synthesize leucine, one finds among every billion cells about 100 individual cells that are unable to do so. If these mutants from different culture tubes are examined, it becomes plain that most differ from one another; in fact, there seem to be about 500 ways by which *E. coli* can lose the ability to synthesize leucine. That may be interesting in itself. More interesting, however, is that of the reverse mutations that restore a bacterium's ability to synthesize leucine, the vast majority fail to correct precisely that lesion which caused the original loss of that ability.

These examples illustrate that the gene content of a species does not have one and only one configuration. It is plastic; a change (mutation) here can be compensated by another one there, and that one (if it has unwanted side effects) can be compensated by still a third change, and, so on, virtually ad infinitum. All this does not imply that the genotype is a fluid sloshing about willy-nilly in response to various environmental challenges. It is plastic in the same sense that glass is plastic; stresses that are too great will shatter glass, whereas steady pressure applied for long periods merely reshapes it. Silly putty is plastic in this same sense.

Figure 9-4 attempts to illustrate the outcome of the genotype's "plasticity." The diagram starts in the near-right corner with species A. A particular gene locus duplicates at an early time, thus giving rise to two genes that make essentially the same protein (all the polypeptide chains that form embryonic, fetal, and adult hemoglobins are extremely similar in their amino acid sequences; they have arisen by such gene duplications). At a later time, still within species A, this locus duplicates once more, thus creating a third copy within the genome.

Within time period 2, species A gives rise to species B. The gene content of the genome of species B contains (at least on a gross scale) all the features contained within A at the time of the split. That includes, for example, the newly arisen polymorphism of protein I. A polymorphism for protein II that arose in species A after B had split off is not, of course, included in B.

Finally, during time period 3, species B gives rise to species C. Existing features within B are passed on to C, but subsequent changes within C lead to differences between it and its parental species (B).

These are the sorts of structural and molecular changes that give rise

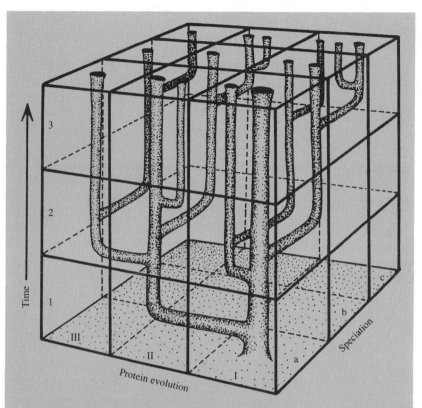

Figure 9-4. The evolution of proteins and the origin of species. Any attempt at this sort of diagram is destined to be grossly oversimplified. Nevertheless, it can help us to understand the occurrence of similar proteins in a given individual and the occurrence of nearly identical proteins in members of different species.

On the axis designated "Protein evolution" three sections are labeled I, II, and III. These represent recognizably different amino acid chains such as the alpha, beta, and delta chains of hemoglobin; such chains are made by different genes. Presumably, the evolution of proteins, in the sense of changing from a protein known by one name to one known by a different name, begins with an abrupt step: the duplication of genetic material (gene duplication). Immediately following gene duplication the proteins synthesized under the control of the two genes must be identical, or very nearly so. Eventually the proteins whose synthesis is controlled by the two duplicate genes diverge in structure because of mutations in the genes themselves. It is the first abrupt step, rather than the gradual divergence following gene duplication, that the figure represents.

Shorter horizontal segments that leave the protein within one or the other of the three classifications represent alterations that do not call for a new designation. For example, nearly all mammals seem to possess several amino acid chains that are recognized as either alpha or beta chains of hemoglobin despite slight differences between them; these chains are made by mutant forms of the genes involved.

Line segments leading back and to the right represent the origin of new species. A line of this sort leads from each protein possessed by the parental species at a given moment; that is, at the time of its origin (except, possibly, for geographic variation), a new species has virtually the same array of proteins as the species from which it has arisen. (Modified from Wallace, 1966.)

to evolution and, at the same time, allow molecular evolutionists to retrace and reconstruct the changes that have occurred through time. The longer the time interval after separation, the greater the dissimilarity of two molecules. The process of evolutionary change begins with the simple genetic differentiation of populations that was considered in Box 1-1. The most elementary evolutionary change is a change in gene frequency. The irreversibility of evolution is a consequence of the huge number of possible solutions that exist for any given problem, either in adapting to new challenges or in readapting to old ones that have returned.

10   The Control of Gene Action

The gene, at least as we last saw it in the previous chapter, is a segment of DNA that, through the intermediation of a molecule of messenger RNA, is responsible for specifying the amino acid–by–amino acid composition of protein molecules (or, more precisely, of their constituent polypeptide chains; a protein molecule such as hemoglobin may be an organized collection of several polypeptide chains). This sequence of events was for many years referred to as the Central Dogma (Figure 10-1): information flows from DNA (an accurately self-replicating molecule) to the messenger RNA (*transcription*) to the protein molecule (*translation*). Transcription is the recopying of information in essentially the same language, as monks did during the Middle Ages. Translation is the transformation of information written in DNA language to information written in protein language— a task entrusted to transfer RNA (tRNA) molecules; this task corresponds to Luther's translation of the Scriptures into German. Figure 10-1 includes a backward step—reverse transcription—in which RNA viruses (such as the AIDS virus) are used as models from which new DNA molecules are generated. The latter can then become incorporated into the host's (DNA) genome. Viral RNA also serves as the template for its own replication.

What genes do is only half the story. *When* they do it is the other half. Both aspects—what and when—of gene action are crucial to normal development and good health. From embryo to adult, all of us have relied on hemoglobin to carry oxygen to our cells and to aid in the disposal of carbon dioxide. As embryos, each of us needed a special (fetal) hemoglobin whose physicochemical characteristics (primarily its affinity for

167

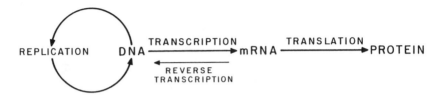

Figure 10-1. The Central Dogma of molecular biology, in which genetic information is shown to flow primarily from DNA (a molecule that can specify its own synthesis) to mRNA (by transcription) to proteins (by translation). Certain enzymes (reverse transcriptases) can synthesize DNA by transcribing RNA. Ordinarily, reverse transcriptases synthesize DNA from RNA viruses; the newly synthesized DNA may then be incorporated within the host's genome, thus allowing the infective virus to arise within a host's cells at any future generation. Experimentalists use reverse transcriptase to generate the DNA that corresponds to an mRNA molecule of interest; by inserting the newly synthesized DNA into a bacterium (via an appropriate plasmid), they can use that bacterium and its descendants to manufacture the corresponding protein. Not shown in this diagram is the replication of RNA viruses, a case in which RNA specifies the synthesis of more viral RNA.

oxygen) differ from the hemoglobin that sustains us as adults. Throughout all this time, hemoglobin has been restricted to our red blood cells. It has not been manufactured in the skin; our nervous system has not become clogged with unwanted hemoglobin. These facts emphasize that the genes responsible for the synthesis of hemoglobin are activated (turned on) in red blood cell precursers—not in the eye, not in the skin, and not in the lining of the digestive system. Furthermore, the turnover from fetal to adult hemoglobin occurs at birth—not several weeks before, not several weeks after.

Patterns based on the localized synthesis of proteins are all about us, but we are so used to seeing them that we do not often recognize the problems they pose. Take, for example, the dark purple little floret in the center of nearly every blossom of Queen Anne's lace, the wild carrot. That floret is so characteristic that F. C. Steward and his colleagues at Cornell University, while developing techniques in which single protoplasts divide and eventually regenerate whole plants, used it to demonstrate the accuracy with which the regeneration process operates. And yet, all wild carrot blossoms do not have a single, deep purple floret in the center. In some cases the purple is deposited around the circumference of the whole flower; in others, the pigment is restricted to a pie-shaped wedge or to centrally located clusters of florets. The shade of purple may vary from flower to flower both within and among plants. Obviously, the formation and deposition of purple pigment occasionally escape normal control. McClintock (1967) recorded such observations, and yet one of her renowned colleagues, years later, asked me, "Why did she waste her time looking at Queen Anne's lace?"

Patterns to some people are not problems; they merely exist. To others they illustrate the exquisite control of gene action. Consider, for example, the red on the head of a red-headed woodpecker. Or the smaller patch of red on the downy woodpecker. Or the spots on a leopard. Or the barred feathers on many birds. Or the tiger or tabby patterns of house cats. Or the mottled kernels on McClintock's Indian corn. A pattern that is recognizable from individual to individual must be rooted in a precise control of gene action.

The *lac* operon

Unraveling the mechanisms by which gene action is controlled has been an extremely complex process. One person who devoted many years of her life to doing exactly that was Barbara McClintock, who wrestled with this problem at Cold Spring Harbor for years. *Zea mays* (Indian corn) was her experimental material. Her efforts proved to be successful. Nevertheless, the first system that was truly understood was in the colon bacillus, *E. coli*. A group of French scientists—François Jacob, Jacques Monod, and E. L. Wollman—was largely responsible for unraveling the control mechanisms in this organism. Once the mechanisms have been explained, the reader may say that they are scarcely more complex than the wiring of a household electrical circuit that permits one to turn off downstairs lights after having gone upstairs to bed. And he or she will be correct. But how many of my readers can diagram the wiring for those electric lights? Worse, perhaps, how many would have recognized the existence in them of a problem demanding solution?

If two small samples of *E. coli* are removed from a culture growing on glucose-containing medium, and one is placed in a tube with glucose-containing medium while the other is placed in a tube whose medium contains the sugar lactose instead, both inocula will grow; the one on glucose starts immediately, the one on lactose starts after an approximately 20-minute delay (Figure 10-2).

If, on the other hand, a tube is used whose medium contains both glucose

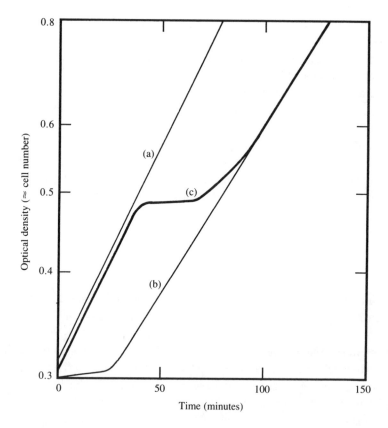

Figure 10-2. The exponential growth of *E. coli* in liquid culture containing minimal medium and a limiting amount of the sugar glucose, lactose, or a glucose-lactose mixture. Cells taken from an initial glucose-fed culture begin growing immediately on glucose (a) but are delayed about 20 minutes when placed in lactose medium (b). A similar delay occurs after the glucose in the glucose-lactose mixture has been exhausted (c). Note that the rate of growth is somewhat slower on lactose than on glucose.

and lactose, the glucose-raised bacteria will grow immediately, increase in number until the glucose is exhausted, pause for about 20 minutes, and then begin growing once more (Figure 10-2). This stepwise growth curve is known as a diauxie curve.

More pieces can be added to the puzzle, even though solving it is not our purpose. The metabolism of lactose requires an enzyme, β-galactosidase, which is formed during each "stationary" period mentioned above and illustrated in Figure 10-2. An enzyme is made under the direction of a gene.

Some mutant bacteria cannot utilize lactose at all; others always synthesize β-galactosidase, and hence their multiplication would not pause under any of the conditions illustrated in Figure 10-2.

By utilizing information such as this and by manipulating *E. coli's* chromosomal DNA in ways not intended by nature, the French workers arrived at the solution illustrated in Figure 10-3. β-Galactosidase is merely one of three enzymes whose structures are encoded in a small segment of bacterial DNA. This region of the chromosome can be referred to as the structural gene (or genes, if all three are to be discussed). Close by and "upstream" (in front of), as the jargon goes, is a second gene (segment of DNA) that specifies a protein called the *repressor*. In the absence of lactose (or of several other small *inducer* molecules), this protein attaches to the DNA just in advance of the structural gene; this attachment region is called the *operator*. As long as the repressor protein is attached to the operator region, the structural genes cannot be transcribed; that is, no messenger RNA can be made. Hence, the cells lack the ability to use lactose.

The repressor molecule can also attach to lactose (more precisely, to a rare chemically variant form of the lactose molecule—allolactose); having done so, the repressor-lactose complex is unable to attach to the operator region of the bacterial chromosome. Under such circumstances, the structural genes can be transcribed and the messenger RNA can lead to the production of β-galactosidase and the other two enzymes. The presence of lactose turns on the machinery needed for its own metabolism; when the

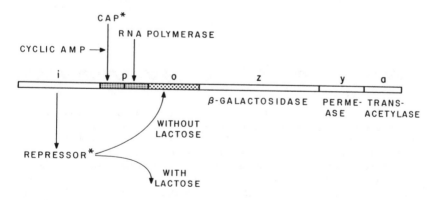

Figure 10-3. The segment of *E. coli's* chromosome concerned with the utilization (or nonutilization) of the sugar lactose. Events concerned with lactose utilization are explained in the text. Asterisks mark regulator proteins.

lactose has been metabolized, the machinery is closed down once more by the lactose-free repressor protein.

What about the events that transpire when a culture whose medium contains both glucose and lactose is inoculated with glucose-grown bacteria? The glucose in this medium activates still another control for the *lac* operon (i.e., the three structural genes under the operator's control). Glucose reduces the amount of cyclic adenosine monophosphate (cAMP) molecules in bacterial cells. As glucose disappears (from both the growth medium and the cells' cytoplasm), this cAMP increases, thus enabling still

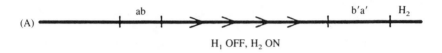

(A)   ab   b'a'   H₂

H₁ OFF, H₂ ON

(B)   ab   b'a'   H₂

H₁ ON, H₂ OFF

(C)

(D)

Figure 10-4. The control or regulation of gene action is achieved in many different ways. In these diagrams, the control of transcription is effected by a "flip-flop" switch. The segment of DNA lying between the reversely oriented sequences (*a b–b' a'*) is inverted by recombination between these flanking sequences (see Figure 8-7); the arrows in diagrams A and B reveal the orientation of this interior segment. If DNA from organisms corresponding to either A or B is heated to 90° it will melt (i.e., the complementary strands will separate); upon recooling they pair up (reanneal) once more as shown in C. A mixture of DNA from organisms corresponding to both A and B that is melted and allowed to reanneal produces many DNA molecules resembling C; however, the mixture also produces molecules resembling D. The unpaired strands in the bubble region have reversed orientations and hence cannot reanneal properly. H₁ and H₂ refer to alternative flagellar proteins in *Salmonella;* only the gene specifying H₂ lies adjacent to the switch.

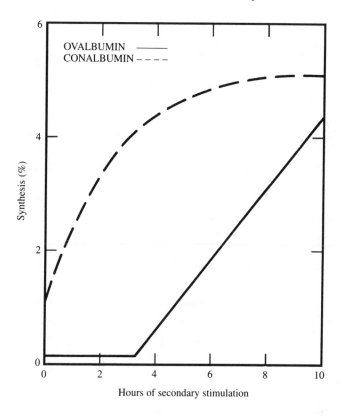

Figure 10-5. Two genes in the cells of a higher organism (the chick) respond differently to the same stimulus (estrogen). When the sex hormone estrogen is injected into a chick, cells in its oviduct synthesize the proteins ovalbumin and conalbumin. The diagram reveals, however, that while the synthesis of conalbumin begins immediately following injection, the synthesis of ovalbumin is delayed by approximately three hours.

another protein (catabolite gene–activator protein, or CAP) to assist the enzyme RNA polymerase to attach to the DNA, upstream of the operator region. Without CAP, the polymerase is unable to transcribe the *lac* operon even though all repressor molecules may be associated with lactose and no repressor molecules are attached to the operator segment.

The control of the *lac* operon—turned on when needed, turned off when not needed—was the first gene control mechanism to be thoroughly understood. The importance of this splendid work was not so much in the details that apply to this one locus in this one bacterium (controls of other gene loci in other organisms differ, sometimes strikingly; see Figures 10-4 and 10-5) but in giving workers a way of viewing the problem, a way of viewing that leads to appropriate experimental analyses. For example, one old textbook (1937) gives the following "equations":

$$G + E_1 \rightarrow 2G + S_1$$
$$G + E_2 \rightarrow 2G + S_2$$

The gene (G) reproducing itself (i.e., giving rise to 2 Gs) in one environment (E_1) must give rise to a substrate (S_1) that differs from the substrate (S_2) that arises when the gene replicates in a second environment (E_2). Globally, this might be true in the sense that a gene product, an enzyme, may react differently in different environments. In particulars, however, the equations are meaningless. The gene product, an enzyme, is not a remnant of the "environment" that remains after gene replication. The two environments (E_1 and E_2) may be used here to represent glucose medium (E_1) and lactose medium (E_2). These environments lead to the production (or its lack) of an enzyme whose chemical composition is specified by a particular region of DNA. The growth of bacteria results not from differing remnants (S_1 and S_2) but from the presence of specific enzymes whose production can be turned off or on, depending upon the environment.

Internalizing the environment

Geneticists are regarded by many as the Presbyterians of biology because they continually stress the importance of the genetic endowment each of us has received from our parents. Especially in the years before 1949, this endowment was seen as the biological equivalent of predestination, the view that God has set each of us in motion much like a mechanical toy with (His) full knowledge of our ultimate fate. What we as human beings do in the meantime is of little or no consequence in determining that fate. Genes, in the eyes of many, possess the same power; do what we might, our fate is sealed when a given sperm (chosen by chance from millions of others) fertilizes a given egg (one of many genetically dissimilar ones shed individually at monthly intervals by our mothers).

The presence of control regions upstream of structural genes allows us to distinguish between two genetic components (Figure 10-6). The first is the chromosomal segment consisting of the structural gene(s) and the adjacent

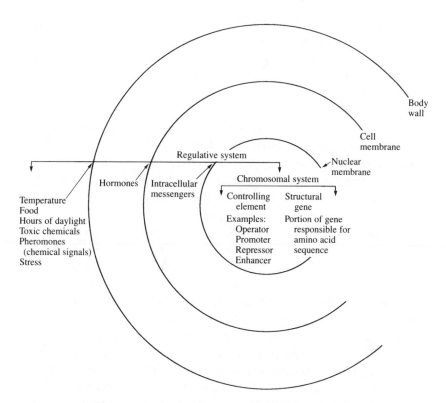

Figure 10-6. Genes whose action is controlled or regulated in some manner can be viewed as being built of two parts. The structural gene and those elements which happen to be closely linked to it correspond to the Mendelian gene—the functional unit—which is generally inherited as an intact block. The closely linked controlling elements, however, also form the terminus of a regulative system whose origin may lie in the world beyond the individual's body wall—in hours of sunlight, degree-days, inches of rainfall, or available nutrients, for example. The integration of heredity and environment—of nature and nurture—within the individual results from the operation of this dual system in which the terminus of one portion forms the initiating (upstream) component of the other.

control regions, which, if not examined too closely, corresponds to the individual factors studied by Mendel. These are the units that provide the Mendelian ratios associated with the inheritance of chromosomal genes.

The second component *terminates* in the control region of the chromosomal gene but arises at any of several possible, more or less remote, locations (Figure 10-6), including the world beyond the individual's body wall.

The *lac* operon I discussed earlier responds to the presence of lactose (and the absence of glucose) in the culture medium. The sensing mechanism depends upon the transport of these sugar molecules across the cell membrane and into the cell, where they react with specific proteins (repressor, CAP).

External stimuli need not penetrate the cell to alter gene action. At times it is sufficient for an "outside" chemical to interact with the exterior portion of a membrane protein, which then alters (for biophysical reasons) the configuration of an internal segment. The change in the configuration of the sensing protein that occurs inside the cell results in biochemical events that impinge on gene control regions—either calling for or preventing the transcription of one or more genes (Figure 10-6).

Without entering into the minutiae of gene-regulating mechanisms— mechanisms that are the focus of much modern genetic research—one can note that the open, ill-defined, even far-out beginning of the gene control system that eventually terminates at the chromosomal component provides the avenue by which the external environment effectively enters the cell and modifies genetic processes occurring within the cell. The predestination that geneticists expounded in the past resulted largely from ignorance. As knowledge concerning gene regulation and gene control becomes more sophisticated, man-made environments will lead to substantive—not trivial—alterations in gene action. As Barbara McClintock once said in introducing a report at Cold Spring Harbor, "With the tools and the knowledge, I could turn a developing snail's egg into an elephant. It is not so much a matter of chemicals because snails and elephants do not differ that much; it is a matter of timing the action of genes."

Where is the gene?

Our search for the gene has now led us, as illustrated in the figure of the *lac* gene in *E. coli* (Figure 10-3), to a circumstance in which one gene (i) specifies a protein (the repressor) whose function is to attach itself to a nearby (in the case of the *lac* gene; more remote for many other genes) region of DNA (the operator), thereby preventing the transcription of the DNA segment (structural gene) that, in the case of the *lac* gene, specifies three enzymes.

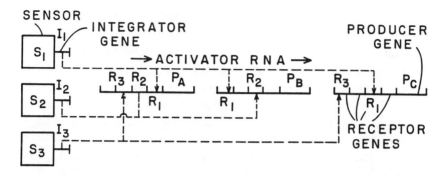

Figure 10-7. One of two gene regulatory models proposed by Britten and Davidson (1969). A given sensor(s) detects an initiating factor (hormone, temperature, light, etc.) and activates a corresponding integrator gene (I) that produces an activator substance (RNA or polypeptide). The activator substance attaches to appropriate receptor genes (R) at various gene loci, thus coordinating the simultaneous activity of many genes. The receptor genes allow transcription of the producer genes (P). Note that the word *gene* as used in the present text is assigned to each collection of Britten and Davidson's receptor genes and their nearby producer gene.

Genes in this account have long since ceased being beads on a string; they *are* the string. They are segments of DNA that grossly resemble every other segment except for the order of their bases and thus the information they encode. Recombination, as Benzer, Yanofsky, and their colleagues showed, occurs within as well as between genes. And why not? Physically, the regions both within and between genes are merely DNA—a double-stranded helix.

How many *kinds* of genes are there? Britten and Davidson (1969), in an often-cited paper, proposed several kinds (Figure 10-7), especially for higher organisms.

Integrator genes. These sense nongenetic signals from either the external or internal environment and, as a result, produce signal proteins (or small RNA fragments).

Receptor genes. The signal proteins made by integrator genes diffuse to and attach to receptor genes (these are equivalent to the operators discussed earlier), which in turn control the producer genes.

Producer genes. These are the structural genes of the preceding paragraphs, the genes that ultimately specify the amino acid sequence in enzymes and other protein molecules.

Perhaps the following definition of *gene* will prove faulty under close scrutiny, but I find it convenient to regard any chromosomal region that exhibits a coordinated relationship such as a structural gene together with its nearby (usually upstream) regulating elements (even those referred to as "genes" by others) as a gene. In higher organisms such a unit exhibits Mendelian inheritance. It is, by definition, a unit of function—not one of mutation or recombination. Control genes that lie on other chromosomes or on the same chromosome but so far removed that recombination is common are referred to as *other* genes, together with all other, unlinked genes. Within the tightly linked region, however, the same structural gene (A) with either one or the other of two dissimilar *cis*-acting (i.e., same DNA strand, or chromosome) essential controls (C_1 and C_2) become, in my view, two alleles of one gene—A_1 and A_2. If the structural gene cannot perform its mission without its control, and the controls of two identical structural genes differ, I say the *genes* differ. In doing so, I virtually ignore the identity of the structural genes. By the same token, I say that sickle-cell and normal hemoglobins differ regardless of the identity that these two hemoglobins exhibit at all but 1 of nearly 300 amino acid sites.

The gene defined by recombination: A second look

Many genes have both a *function* (specifying the structure of a protein) and a *proper time and place* for carrying out that function. The *lac* operon is a good example: The main function of that operon is the synthesis of β-galactosidase (and two other enzymes); that is, it specifies the sequence of the amino acids that constitute that enzyme. The proper time to produce β-galactosidase is (1) when lactose is present in the surrounding medium and (2) when glucose is not present. Restricting the production of β-galactosidase to the proper times is the function of the upstream control region of DNA: the operator, the promoter, and (at other gene loci) still other controlling elements.

Higher organisms all have genetic controls; some are so commonplace that we scarcely notice them. Hemoglobin is made only by cells destined to become red blood cells; even the "sibling" cells that are destined to become white blood cells are exempt. Pepsin, trypsin, and other digestive enzymes are synthesized only by cells lining the digestive tract. If the cells lining tear ducts or constituting nasal mucosa were to synthesize those enzymes or secrete hydrochloric acid as cells lining the stomach wall do, it would be painful indeed. The leopard's spots, the tiger's and zebra's stripes, and the complex color patterns of many birds are the result of many precise genetic control mechanisms; these controls are responsible for the synthesis of protein in the right cells at the right time.

Certain enzymes synthesized by *Drosophila melanogaster* (and which can be detected by electrophoresis) are made neither uniformly throughout the development of the fly (egg, larva, pupa, and imago) nor indiscriminately among the various tissues (see Figure 10-8). If a segment of DNA spanning the entire gene from control region to structural gene is removed from flies belonging to one *Drosophila* species and is inserted into flies of a different *Drosophila* species, the details of enzyme synthesis may change (Brady and Richmond, 1990). The gene product may appear as expected in some tissues but not in others. Either the signals calling for its presence are not being elicited or the appropriate signal receptors are lacking.

Just where these facts lead us can be seen if we draw the following diagram:

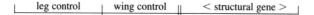

| leg control | wing control || < structural gene > |

and then recall the nature (or logic) of a standard genetic test: that test locates *differences* between two chromosomes (or DNA strands). The geneticist studies the segregation or separation of *dissimilar* elements.

If, with respect to the diagram above, two forms (normal and mutant) exist at the leg control region, the consequent malfunctioning of the entire locus may lead to abnormally formed legs. Alternatively, a mutation within the wing control region may lead to abnormal wings. An observer may interpret these as two unrelated events: there appears to be a "leg" gene and a "wing" gene. What, then, will genetic analyses reveal? The interpretation will be verified because the structural gene (as the situation has been described) is unchanged. The two "genes" will be located by recombinational tests as being (1) closely linked and (2) with the "leg" gene lying to the left of the "wing" gene. If other landmarks were available, a molecular analysis of this region of DNA might reveal that this region which ostensibly embraces two "genes" is really too short to contain even one gene. Results of precisely this sort have been reported (Hirsh, 1989) for the dopa decarboxylase gene in *Drosophila* (Figure 10-9).

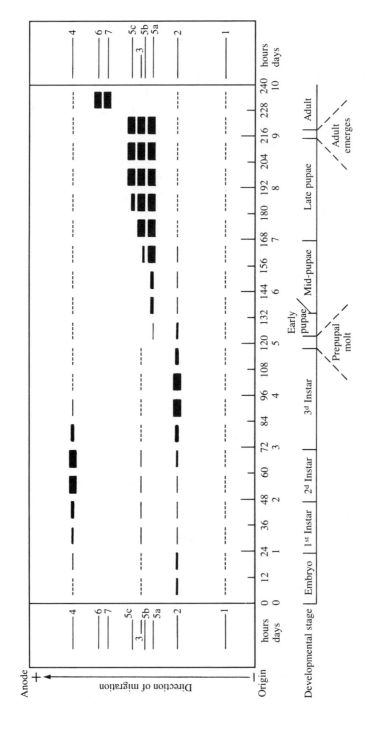

Figure 10-8. Patterns of alkaline phosphatase activity in *Drosophila melanogaster* during its development from embryo to adult. At the left is a diagram of an electrophoretic gel showing the positions of nine recognizable enzyme-generated bands. The horizontal scales represent time (hours, days) and developmental stages. Degrees of activity (i.e., amounts of enzyme) are represented by lines of five widths: dashed lines represent trace activity and the thickest lines represent maximum activity. (Modified from Sena, 1966.)

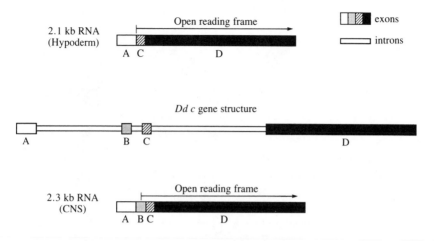

Figure 10-9. The dopa decarboxylase gene in *Drosophila melanogaster* is required for the proper hardening and pigmentation of the cuticle (hypoderm) and for the synthesis of neurotransmitters in a few selected neurons of the central nervous system. In preparation for translation, the mRNAs are processed in two different ways, with the result that the proteins synthesized, although similar, differ in the two tissues. Note that following standard genetic tests, mutational changes in the DNA at regions A, B, and C could be interpreted as indicating the presence of three closely linked genes. (intron = a portion of transcribed mRNA that is removed before the mRNA becomes functional; exon = a portion of transcribed mRNA that, when spliced to other exons, forms the functional mRNA [*exon* = *expressed*]; kb = kilobase [1000 nitrogenous bases]; CNS = central nervous system.) (Modified from J. Hirsch, *Developmental Genetics* 10:232–238. Copyright © 1989 by Alan R. Liss. Reprinted by permission of Wiley-Liss, a division of John Wiley and Sons, Inc.)

One of the more interesting studies of these "subgenes" was carried out by Soviet geneticists during the late 1920s. The normal *Drosophila* fly (Figure 10-10) has numerous (nearly 30) major bristles arranged on its head and thorax in characteristic positions. Many mutant genes "remove" various bristles; more accurately, these genes interfere with the development of these bristles, which in itself is a complex biological feat involving originally a single cell for each.

Scute, a gene locus near the distal end of the X chromosome of *Drosophila melanogaster,* is especially important (for unknown reasons) in governing the normal development of bristles. Some mutants at this locus remove nearly all bristles from the affected flies. Others remove only a few—one removes them here, another removes them there; one removes bristles at only one or two paired sites (the left and right sides of a fly are symmetrical), another at several sites.

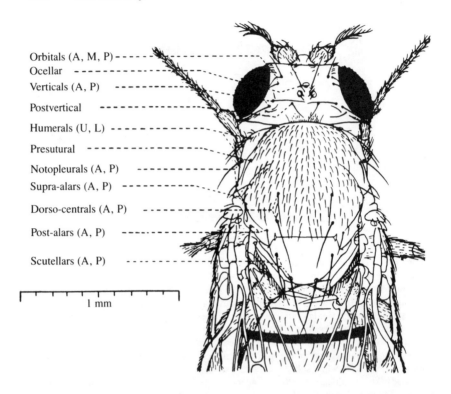

Orbitals (A, M, P) - - - - - -
Ocellar - - - - - - - - - - -
Verticals (A, P) - - - - - - -
Postvertical   - - - - - - - -
Humerals (U, L) - - - - - - -
Presutural - - - - - - - - -
Notopleurals (A, P) - - - - - -
Supra-alars (A, P) - - - - - - -
Dorso-centrals (A, P) - - - - - -
Post-alars (A, P)   - - - - - - -
Scutellars (A, P)   - - - - - - -

1 mm

Figure 10-10. The head and thorax of *Drosophila melanogaster,* indicating the position of 20 major setae (bristles). Each bristle starts as a single cell among hundreds or thousands of apparently identical cells that eventually form the hypodermis of the developing adult. The "bristle" cell divides, after which one daughter cell grows rapidly (1000-fold within 24 hours) and (within 36 hours) manufactures the bristle. The smaller cell forms a socket around the base of the bristle; a nerve cell also makes contact with these two and, together, the three constitute a small but efficient sensory unit. Different genera and families of flies have bristles at different locations on their bodies, but within a genus—certainly within a species—these locations are extremely precise. (A, M, P, U, and L are anterior, medial, posterior, upper, and lower, respectively.)

The Soviet geneticists claimed that the names of these bristles could be arranged in an order such that each mutant removed only those bristles whose names were next to one another. (This order, incidentally, bore no relation to the actual distribution of bristles on a fly!) Furthermore, if recombination occurred between *scute* alleles that removed bristles at opposite ends of this array, one could obtain recombinant mutant alleles that removed nearly all bristles.

The conclusion reached as the result of these experiments, much like that of Charles Yanofsky and his colleagues, was that there existed a *colinearity* between the bristles on the fly (these having been arranged in a particular order) and subgenes at the *scute* locus: each subgene was responsible for a given bristle. Now, it had been known from the early days of genetics that aspects of the phenotype that can be enumerated (like fingers and toes) are not the result of particular genes: there are no genes for little fingers, for ring fingers, for index fingers, for thumbs, for noses, and for other specific parts of the body. Normal development of digits, extremities, vertebrae, teeth, and other anatomical features requires the functional interactions of many genes. A mutant gene may add an extra finger (polydactyly), but that does not mean that the normal gene ordinarily removes an otherwise extra finger. To cite a more familiar analogy: dirt in the gas line may cause an automobile engine to stop running; that does not mean that a clean gas line guarantees the proper functioning of a car's engine. One or two words may explain why an engine stops; several volumes are needed to explain why it runs.

I mention the Soviet view (during the 1920s) about subgenes at the *scute* locus because, in its day, it was a sensational claim. A. H. Sturtevant (of genetic map fame) and Jack Schultz, two American geneticists, disputed the Soviets' interpretation but confirmed what the Soviet scientists had said about an order (or sequence) of bristles such that those removed by any one *scute* allele were neighbors on the assembled list. H. J. Muller is reported to have said much later that the *scute* story is still not understood. The subgenes might in fact represent control sites like those I described earlier. A thorough molecular analysis will be needed to interpret all the data obtained by classic genetic methods. (A molecular analysis has indeed been undertaken but, to my knowledge, not summarized for the nonprofessional reader; see, for example, Ruiz-Gomez and Modolell, 1987.) Classical methods, we might recall, merely identify places at which two chromosomes (or DNA strands) differ; the function that seems to reside at the identified site may involve the cooperative interaction of neighboring elements that occupy a much more extensive chromosomal region.

11 A Potpourri of
Gene Manipulations

A striking feature of biochemistry during the past three decades has been the advances its practitioners have made in the manipulation and subsequent analysis of macromolecules. In large measure this development followed the intrusion of genetics into an area of research that, until then, had concentrated largely on the identification of the enzymes involved in ever more obscure metabolic pathways. To study DNA and proteins in detail requires techniques that leave large, fragile molecules intact. But if these large molecules must be broken into smaller fragments, the breaks must be at specific sites, and each technique has been designed to deal with a different set of sites. For the most part, these techniques rely upon the use of special enzymes that are themselves macromolecules.

Circular DNA

Until now the discussion of DNA has emphasized its linear structure: 20 Å wide and terribly long, as one person expressed it. The same person (Frank Stahl) was also instrumental in discovering the circular structure of bacterial chromosomes. Studies of the linkage of mutant genes in *E. coli* had yielded results which, although consistent in many respects, were strangely inconsistent. For example, a study of five mutant genes may have suggested the following linkage relationships:

1. *a-b-c-d-e*
2. *d-e-a-b-c*
3. *b-c-d-e-a*

The interpretation of such patterns comes more easily to magicians and inveterate card players than to others, perhaps, because a deck of cards is logically a circle. No matter what the order of cards in a well-shuffled deck may be, a single cut of the deck does not alter that order. Cutting merely places the bottom card on top of the top card, thus mending one break in the circle while opening the circle at a new location. Examination of the three linkage relationships shown above reveals that a circular chromosome containing the five mutant genes shown below

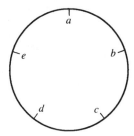

would give each of the observed sequences if the circle were cut (1) between *e* and *a*, (2) between *c* and *d*, and (3) between *a* and *b*. An electron micrograph of the circular chromosomes of *E. coli*, in the process of replicating, is shown in Figure 11-1. The circular structure of chromosomes is a characteristic of prokaryotes (cells without distinct nuclei), but not of eukaryotes (cells in which a distinct nucleus is evident).

Chloroplast and mitochondrial DNA

Earlier I mentioned that the Feulgen test revealed that DNA is restricted to the nucleus; however, a brief further comment noted the physical limitations of the light microscope. Now is the time to point out that two cytoplasmic organelles—chloroplasts in plant cells and mitochondria in

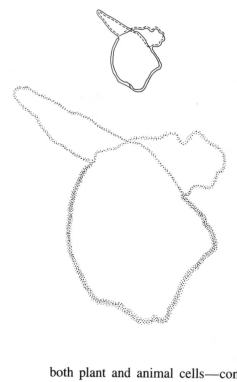

Figure 11-1. Diagram based on a photograph of a dividing *E. coli* chromosome. When provided with tritium-labeled thymidine (a molecular unit consisting of deoxypentose and thymine) bacteria synthesize radioactive DNA. Thus, chromosomes obtained from bacterial cells that have been removed to a nonradioactive medium reveal their presence by exposing a super-imposed, radiation-sensitive film to beta radiation. The parental portion of the chromosome reveals its presence by the more numerous silver granules in the film; both strands of DNA in this portion are radioactive. The daughter portions possess only one radioactive strand, and thus generate fewer silver granules in the film. In the smaller drawing, parental DNA strands are shown as solid lines; the newly synthesized (nonradio-active) strands are shown as dashed lines. (Redrawn from Cairns, 1963, by permission of Cold Spring Harbor Laboratory.)

both plant and animal cells—contain small amounts of circular DNA. Chloroplasts had been recognized as self-replicating units within plant cells, and theorists such as H. J. Muller suspected that DNA must therefore be present, because the ability to self-replicate requires access to information concerning protein synthesis and timing. Those are the functions of DNA.

Cellular organelles are not self-sufficient; they are unable to function independently of the cell's nuclear genes. In higher plants, for example, many nuclear genes help control the synthesis of chlorophyll, the green pigment contained within chloroplasts that is the hallmark of most plants. Mutant forms of these nuclear genes exhibit typical Mendelian inheritance patterns; mutant forms of chloroplast genes are transmitted primarily through the cytoplasm of the egg. Male gametes have very little cytoplasm; animal spermatozoa, for example, do not transmit mitochondria from the father to his offspring.

Organism	Codon	Usual Meaning	Meaning in Mitochondrion
Common Mammal	UGA AG$_G^A$	termination arginine	tryptophane termination
Mammal Fruit fly Yeast	AUA AUA AUA	isoleucine isoleucine isoleucine	Met (initiation) Met (initiation) Met (elongation)
Yeast Fruit fly	CUA AGA	leucine arginine	threonine serine

Figure 11-2. The genetic code is virtually universal in the sense that messenger RNA specifies the same sequence of amino acids whether the cellular machinery is bacterial, plant, or animal. Differences in the meanings of some codons are found, however, when the mitochondrial genetic code is compared with the "universal" one. Whether these differences reflect ancient ones tracing from diverse origins of genetic codes or useful differences that have arisen following the adoption of symbiosis as a way of life is not known.

The DNA of cytoplasmic organelles is unlike that of the cell's nucleus in several respects. I have implied up to now that the genetic code is universal, but several triplet codons in mitochondria have meanings that differ substantially from the nuclear code (see Figure 11-2). Furthermore, chloroplasts and mitochondria react to many antibiotics in a manner more closely resembling that of bacteria than that of the cells of higher organisms within which these organelles are found. For these and other reasons, some people postulate that early in the history of life on earth, certain symbiotic relationships were established among already free-living forms. These symbiotic combinations (much like those of today's lichens [algal-fungal symbiotic combinations]) are so ancient and the reliance of one cell type on the other is so complete that they are perceived as a *single* form of life. Carl R. Woese (1984) has done much to unravel the complexities arising from the intimate intermingling of early life forms (Figure 11-3).

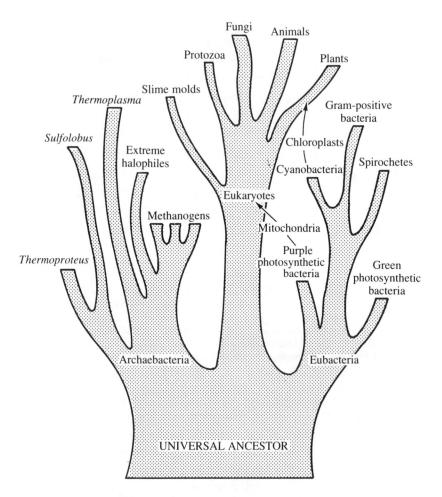

Figure 11-3. An attempt at grouping all living organisms into a single phylogenetic tree. The most interesting aspects of this tree are the solid black arrows that designate the adoption of symbiosis as an enduring way of life for mitochondria and chloroplasts. Millions of years in the future similar arrows may be needed to represent the descendants of what are today called lichens—algae and fungi that live symbiotically. (Modified with permission from Carl R. Woese, *The Origin of Life,* Carolina Biology Reader Series, no. 13. Copyright 1984, Carolina Biological Supply Co., Burlington, N.C.)

Plasmids

Early studies of antibiotic resistance among the bacteria responsible for diarrhea revealed that this resistance spread among bacteria of different species, or even different genera, at an alarming rate. The explanation lay in small, circular bits of DNA, each large enough to specify only a few protein molecules (that is, each containing only a few genes), known as plasmids. They frequently carry genes that synthesize enzymes that destroy various antibiotics. Although the use of antibiotics in modern medicine dates from the World War II era, bacteria have been exposed to inhibitory substances secreted by *Penicillium, Streptomyces,* and other organisms (including other bacteria) for eons.

Plasmids need not replicate in synchrony with bacterial cell division; unless their replication is inhibited by intracellular reactions, their number within an individual cell can become large. Furthermore, they can be transferred from one cell to another. Thus, when exposed to an antibiotic, an entire culture of bacteria can acquire resistance without the need for cell division, many deaths, and the replacement of the original sensitive population by mutant, resistant individuals. Resistance to an antibiotic can spread from one bacterium to another (by plasmid transfer) much like an infectious disease. The situation is actually much more complex than mere plasmid sharing among the bacterial inhabitants of one's intestines; nearly all bacteria, whatever their mode of life, have access to these plasmids (and the information they carry— much of which bears on antibiotic resistance), as Figure 11-4 illustrates.

Cruciform DNA

DNA, the genetic material for most living things (RNA serves for the remainder), has a characteristic structure. Because it is the genetic material— the chemical that transmits information on how and when to synthesize both structural and enzymatic proteins from generation to generation— DNA imposes some constraints. In civil engineering and architecture one refers to strengths of materials; strengths of materials have influenced the shape of buildings from the pyramids to the Parthenon to the grand cathe-

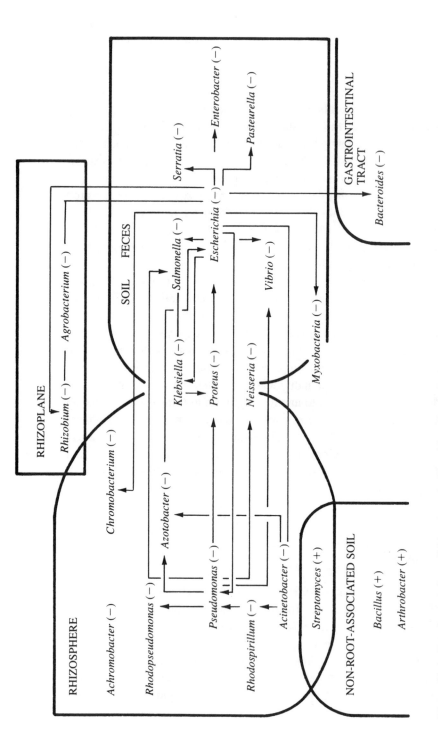

Figure 11-4. Genetic networks available to bacteria. Plasmids and bacteriophage can be exchanged among bacteria belonging to different genera; hence, changes (such as the acquisition of resistance to an antibiotic) that occur in one group can be transmitted through the network's circuits to organisms that superficially seem not to be involved. Of special importance for human beings is the number of lines leading into and out of our large intestines (soil–feces). (Modified from Davey and Reanney, 1980, by permission of *Evolutionary Biology*.)

drals of Europe to the modern skyscraper. The physical characteristics of DNA have similarly influenced its use as genetic material.

The two strands of DNA can be separated by thermal energy. The two hydrogen bonds that hold adenine to thymine (A=T) break down at a lower temperature than do the three hydrogen bonds that bind guanine to cytosine (G≡C). This fact can be used to reveal A=T–rich regions of DNA (by subjecting DNA to heat). In a test tube at approximately 90° and low salt concentration, all double-stranded DNA becomes single stranded: all bonds are melted. These single strands come together again and reanneal with complementary pairing as the temperature is lowered once more. The fragility of DNA with regard to heat was thought to account for the synthesis in virtually all forms of life—from bacteria to the more complex "higher" organisms—of heat-shock proteins. If cells are rapidly exposed to a brief heat shock, most metabolic functions cease and protein synthesis is devoted almost entirely to the production of these special proteins. Some of these move rapidly into the nucleus, while others remain in the cytoplasm. The consensus is that these heat-shock (or stress-induced) proteins stabilize the configuration of other proteins; for that reason, they are also called *chaperonins*. Microorganisms protect their DNA by relying on high ionic concentrations and as high a proportion (70% in some instances) of G≡C bonds as possible. A few bacteria survive and reproduce under high hydrostatic pressure at temperatures exceeding 100°C (the boiling point of water at sea level).

Because of its structure, the following reverse-repeat (palindromic) structure of DNA is of special interest to molecular biologists:

$$\downarrow$$
ATTGCTACGATCGTAGCAAT
TAACGATGCTAGCATCGTTA
$$\uparrow$$

Recalling that the orientation of the two strands of DNA is in opposite directions (the strands are antiparallel), one can see that the upper and lower strands of this segment are identical. They are also symmetrical about the point designated by the arrows. Those who have played cat's cradle can imagine hooking their little fingers between the two strands of DNA at the arrows and pulling outward. The result would be:

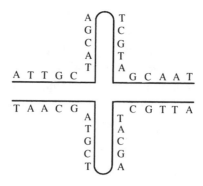

As the horizontal strands unwind and separate, each cruciform double strand forms a double helix. Virtually no energy is required to pull the strands out as illustrated here, or to allow them to go back to their standard, double-stranded linear form. Many people believe that cruciform structures of the sort shown here serve a purpose, but precisely what that purpose may be remains unclear. Indeed, these reverse orientations within DNA may be more important with respect to the synthesis of RNA that can form double-stranded regions (thus stabilizing the structure of the RNA molecule) than to the DNA molecule itself.

Restriction enzymes

An examination of the point of symmetry of the palindromic DNA molecule shown above reveals the following bases:

CGATCG

GCTAGC

or, if we restrict our interest to a segment only four bases long:

GATC

CTAG

These small repeat areas have proven to be especially important to molecular biologists. First of all, with only four bases involved in their construction, there are only sixteen possible configurations.* Second, for each possible construction, there seems to be an enzyme (restriction enzyme) that can cut the DNA molecule at that palindromic spot:

TTCAGATC   GGATC
AAGT   CTAGCCTAG

The enzyme cuts each of the two strands in a corresponding site (beyond the final C of the palindrome), thus leaving the otherwise double-stranded end with a single-stranded projection—a "sticky" end. "Sticky" because the GATC at one end can re-pair with the CTAG at the other. If they were to pair again, a second enzyme (a ligase) would quickly repair the gaps in the two strands and restore the intact nature of the DNA. (Some restriction enzymes cut palindromes cleanly in two, creating DNA fragments that possess blunt ends; these, too, can rejoin—any blunt end to any other one—but only with considerable difficulty.)

This account, like many in earlier chapters, could lead into an elaborate presentation of the techniques of molecular biology and genetic engineering. That, however, is not the intent of this book. The following illustration of how one can insert a foreign gene into *E. coli,* thereby causing this bacterium to make the "foreign" gene product, must suffice.

Suppose we have a plasmid that generally resides in *E. coli.* It confers resistance (K^r) to kanamycin (an antibiotic) and has a single site at which a given restriction enzyme can cut it. That plasmid might be represented as a circle:

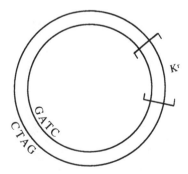

*The leftmost pair may be G-C, C-G, A-T, or T-A, and the same four possibilities exist for the next pair; the remaining two pairs are then fixed because the four pairs form a palindrome.

Suppose, too, that the gene that makes a certain polypeptide hormone in mice is flanked by the same restriction sites:

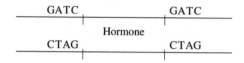

Exposure of a mixture of these two DNAs to the restriction enzyme leads to the following:

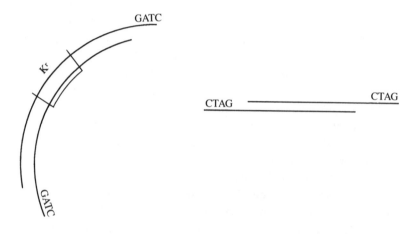

Because of their corresponding sticky ends, these two DNA fragments can unite and, in the presence of a ligase, form a large plasmid.

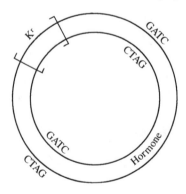

Finally, exposure of *E. coli* that are sensitive to kanamycin to this mixture will result in the infection of certain individual bacteria by plasmids. If these originally sensitive bacteria are now placed on a medium containing kanamycin, only a few cells form colonies—these are the individuals that have gained the kanamycin-resistant plasmids. These colonies can be maintained as thriving cultures and further tested to see if the infecting plasmid has also brought in the hormone-producing gene (if so, the plasmid will be unexpectedly large). The cultures that contain the hormone gene can now be used to produce the mouse hormone in commercial quantities.

A score of details have been omitted in this account. The plasmid must have a promoter region before the *E. coli* cell will transcribe anything from the plasmid. All introns, regions normally excised from the RNA of higher organisms (see Figure 10-9), must be removed from the mouse gene (the mouse DNA used in this procedure would most likely have been obtained by the use of reverse transcriptase and the proper messenger RNA from the mouse) so that *E. coli* can synthesize the hormone correctly. Plasmids whose number in a bacterial cell is not under strict control are more valuable for commercial use than other plasmids. The outcome, however, is that the small colon bacterium, *E. coli,* can be used as an extremely efficient factory for producing substances—proteins—that are normally made in smaller quantities by larger, less convenient, and more expensive animals. The factory does not run without energy, however. One molecule of glucose "burned" to yield CO_2 and H_2O makes available 36 molecules of adenosine triphosphate (ATP), the molecule that serves as the medium of exchange for energy in living things. Every amino acid that is inserted into a polypeptide chain requires 6 molecules of ATP. Thus, 1 kilogram of glucose is required (above the normal maintenance needs of *E. coli*) to manufacture 4.75 kilograms of hormone. At the current costs of glucose and of hormones, however, this exchange is a bargain!

12   Know Thy(DNA)self!

Shortly after Watson and Crick proposed their model of DNA structure, and as it became clearer that the sequence of base pairs along the length of the DNA molecule was related in some point-by-point fashion to the sequence of amino acids in proteins, geneticists (especially those who were ignorant of physical realities) began dreaming of *the* giant microscope— the microscope that would magnify DNA molecules so much that an observer could simply jot down the sequence of base pairs in the molecule under observation. With its powers of magnification, the giant microscope would do for DNA workers what the giant salivary gland chromosomes had done for light microscopists. That was the fantasy of many. And it *was* fantasy.

The ability to "read" DNA bases and jot them down on paper (or punch them into a computer) as rapidly as one's fingers can move came about not through a physical instrument but through a technique—or combination of techniques. These are described in some detail in Box 12-1. Within the narrative portion of the text, it is enough to refer the reader to Figure 12-1, which shows an autoradiograph of high-resolution polyacrylamide gel within which radioactively labeled DNA has been electrophoresed. Fragments of different sizes and charges (obtained as explained in Box 12-1) migrate different distances. Because of the chemicals used in creating these fragments, each vertical column terminates in the purine or pyrimidine indicated at the top. To determine the sequence of bases in the original molecule of DNA, one merely starts at the bottom and reads (or writes

196

Figure 12-1. A "sequencing" film from which the sequence of bases in a given fragment of DNA can be determined. The method for generating such films is described in the text (and Box 12-1); here, it is sufficient to point out that, starting at the bottom, this film reads CAAAAAACGG. . . . The technique for making such films has been standardized so that all graduate students in genetics, biochemistry, and molecular biology now learn it as a routine exercise.

down) the letter heading the column for all columns in succession. The sequence begins as follows:

CAAAAAACGGACCGGGTGTA

I leave it to the reader to check the accuracy of what I have recorded as well as to continue noting more of the sequence. Clearly, the sequence of bases in a DNA molecule can be determined experimentally.

The "molecules" of DNA, except for a few small plasmids, are themselves fragments of much longer strands. Recall that a chromosome consists of a single, extremely long fiber of DNA. How, then, can the extremely long sequence be deduced from preparations such as the one illustrated in Figure 12-1? Rather easily: the terminal arrangements of bases found at the ends of one short fragment can be matched with identical arrangements at the ends of two other fragments, and the three sequences can be reconstructed as a single unit. Figure 12-2 shows a sequence of 2250 base pairs that specify 750 amino acids (still an incomplete sequence because a stop codon has not yet been encountered) that was reconstructed by sequencing 16 overlapping DNA segments. The practice of using overlapping patterns has been long used by archaeologists who date ancient

```
TGA AAT ATG GAG AAT AGT CTT AGA TGT GTT TGG GTA CCC AAG CTG GCT TTT GTA CTC TTC GGA GCT TCC TTG CTC AGC GCG CAT CTT CAA   bp84
 *   M   E   N   S   L   R   C   V   W   V   P   K   L   A   F   V   L   F   G   A   S   L   L   S   A   H   L   Q        aa28

GTA ACC GGT TTT CAA ATT AAA GCT TTC ACA GCA CTG CGC TTC CTC TCA GAA CCT TCT GAT GCC GTC ACA ATG CGG GGA GGA AAT GTC CTC   bp174
 V   T   G   F   Q   I   K   A   F   T   A   L   R   F   L   S   E   P   S   D   A   V   T   M   R   G   G   N   V   L    aa58

CTC GAC TGC TCC GCG GAG TCC GGA CGG GGA GTT CCA GTG ATC AAG TGG AAG AAA GAT GGC ATT CAT CTG GCC TTG GGA ATG GAT GAA AGG   bp264
 L   D   C   S   A   E   S   D   R   G   V   P   V   I   K   W   K   K   D   G   I   H   L   A   L   G   M   D   E   R    aa88

AAG CAG CAA CTT TCA AAT GGG TCT CTG CTG ATA CAA AAC ATA CTT CAT TCC AGA CAC CAC AAG CCA GAT GAG GGA CTT TAC CAA TGT GAG   bp354
 K   Q   Q   L   S   N   G   S   L   L   I   Q   N   I   L   H   S   R   H   H   K   P   D   E   G   L   Y   Q   C   E   aa118

GCA TCT TTA GGA GAT TCT GGC TCA ATT ATT AGT CGG ACA GCA AAA GTT GCA GTA GCA GGA CCA CTG AGG TTC CTT TCA CAG ACA GAA TCT   bp444
 A   S   L   G   D   S   G   S   I   I   S   R   T   A   K   V   A   V   A   G   P   L   R   F   L   S   Q   T   E   S   aa148

GTC ACA GCC TTC ATG GGA GAC ACA GTG CTA CTC AAG TGT GAA GTC ATT GGG GAG CCC ATG CCA ACA ATC CAC TGG CAG AAG AAC CAA CAA   bp534
 V   T   A   F   M   G   D   T   V   L   L   K   C   E   V   I   G   E   P   M   P   T   I   H   W   Q   K   N   Q   Q   aa178

GAC CTG ACT CCA ATC CCA GGT GAC TCC CGA GTG GTG GTC TTG CCC TCT GGA GCA TTG CAG ATC AGC CGA CTC CAA CCG GGG GAC ATT GGA   bp624
 D   L   T   P   I   P   G   D   S   R   V   V   V   L   P   S   G   A   L   Q   I   S   R   L   Q   P   G   D   I   G   aa208

ATT TAC CGA TGC TCA GCT CGA AAT CCA GCC AGC TCA AGA ACA GGA AAT GAA GCA GAA GTC AGA ATT TTA TCA GAT CCA GGA CTG CAT AGA   bp714
 I   Y   R   C   S   A   R   N   P   A   S   S   R   T   G   N   E   A   E   V   R   I   L   S   D   P   G   L   H   R   aa238

CAG CTG TAT TTT CTG CAA CGA CCA TCC AAT GTA GTA GCC ATT GAA GGA AAA GAT GCA GTC CTG GAA TGT TGT GTT TCT GGC TAT CCT CCA   bp804
 Q   L   Y   F   L   Q   R   P   S   N   V   V   A   I   E   G   K   D   A   V   L   E   C   C   V   S   G   Y   P   P   aa268

CCA AGT TTT ACC TGG TTA CGA GGC GAG GAA GTC ATC CAA CTC AGG TCT AAA AAG TAT TCT TTA TTG GGT GGA AGC AAC TTG CTT ATC TCC   bp894
 P   S   F   T   W   L   R   G   E   E   V   I   Q   L   R   S   K   K   Y   S   L   L   G   G   S   N   L   L   I   S   aa298

AAT GTG ACA GAT GAT GAC AGT GGA ATG TAT ACC TGT GTT GTC ACA TAT AAA AAT GAG AAT ATT AGT GCC TCT GCA GAG CTC ACA GTC TTG   bp984
 N   V   T   D   D   D   S   G   M   Y   T   C   V   V   T   Y   K   N   E   N   I   S   A   S   A   E   L   T   V   L   aa328

GTT CCG CCA TGG TTT TTA AAT CAT CCT TCC AAC CTG TAT GCC TAT GAA AGC ATG GAT ATT GAG TTT GAA TGT ACA GTC TCT GGA AAG CCT   bp107
 V   P   P   W   F   L   N   H   P   S   N   L   Y   A   Y   E   S   M   D   I   E   F   E   C   T   V   S   G   K   P   aa358

GTG CCC ACT GTG AAT TGG ATG AAG AAT GGA GAT GTG GTC ATT CCT AGT GAT TAT TTT CAG ATA GTG GGA GGA AGC AAC TTA CGA ATA CTT   bp116
 V   P   T   V   N   W   M   K   N   G   D   V   V   I   P   S   D   Y   F   Q   I   V   G   G   S   N   L   R   I   L   aa388

GGG GTG GTG AAG TCA GAT GAA GGC TTT TAT CAA TGT GTG GCT GAA AAT GAG GCT GGA AAT GCC CAG ACC AGT GCA CAG CTC ATT GTC CCT   bp125
 G   V   V   K   S   D   E   G   F   Y   Q   C   V   A   E   N   E   A   G   N   A   Q   T   S   A   Q   L   I   V   P   aa418

AAG CCT GCA ATC CCA AGC TCC AGT GTC CTC CCT TCG GCT CCC AGA GAT GTG GTC CCT GTC TTG GTT TCC AGC CGA TTT GTC CGT CTC AGC   bp134
 K   P   A   I   P   S   S   S   V   L   P   S   A   P   R   D   V   V   P   V   L   V   S   S   R   F   V   R   L   S   aa448

TGG CGC CCA CCT GCA GAA GCG AAA AGG AAC ATT CAA ACT TTC ACG GTC TTT TTC TCC AGA GAG GGT GAC AAC AGG GAA CGA GCA TTG AAT   bp143
 W   R   P   P   A   E   A   K   G   N   I   Q   T   F   T   V   F   F   S   R   E   G   D   N   R   E   R   A   L   N   aa478

ACA ACA CAG CCT GGG TCC CTT CAG CTC ACT GTG GGA AAC CTG AAG CCA GAA GCC ATG TAC ACC TTT CGA GTT GTG GCT TAC AAT GAA TGG   bp152
 T   T   Q   P   G   S   L   Q   L   T   V   G   N   L   K   P   E   A   M   Y   T   F   R   V   V   A   Y   N   E   W   aa508

GGA CCG GGA GAG AGT TCT CAA CCC ATC AAG GTG GCC ACA CAG CCT GAG TTG CAA GTT CCA GGG CCA GTA GAA AAC CTG CAA GCT GTA TGT   bp161
 G   P   G   E   S   S   Q   P   I   K   V   A   T   Q   P   E   L   Q   V   P   G   P   V   E   N   L   Q   A   V   S   aa538

ACC TCA CCT ACC TCA ATT CTT ATT ATT ACC TGG GAA CCC CCT GCC TAT GCA AAC GGT CCA GTC CAA GGT TAC AGA TTG TTC TGC ACT GAG GTG   bp170
 T   S   P   T   S   I   L   I   I   T   W   E   P   P   A   Y   A   N   G   P   V   Q   G   Y   R   L   F   C   T   E   V   aa568

TCC ACA GGA AAG GAA CAG AAT ATA GAG GTT GAT GGA CTA TCT TAT AAA CTG GAA GGC CTG AAA AAG TTC ACC GAA TAT AGT CTT CGA TTC   bp179
 S   T   G   K   E   Q   N   I   E   V   D   G   L   S   Y   K   L   E   G   L   K   K   F   T   E   Y   S   L   R   F   aa598

TTA GCT TAT AAT CGC TAT GGT CCG GGC GTC TCT ACT GAT GAT ATA ACA GTG GTT ACA CTT TCT GAC GTG CCA AGT GCC CCG CCT CAG AAC   bp188
 L   A   Y   N   R   Y   G   P   G   V   S   T   D   D   I   T   V   V   T   L   S   D   V   P   S   A   P   P   Q   N   aa628

GTC TCC CTG GAA GTG GTC AAT TCA AGA AGT ATC AAA GTT AGC TGG CTG CCT CCT CCA TCA GGA ACA CAA AAT GGA TTT ATT ACC GGC TAT   bp197
 V   S   L   E   V   V   N   S   R   S   I   K   V   S   W   L   P   P   P   S   G   T   Q   N   G   F   I   T   G   Y   aa658

AAA ATT CGA CAC AGA AAG ACG ACC CGC AGG GGT GAG ATG GAA ACA CTG GAG CCA AAC AAC CTC TGG TAC CTA TTC ACA GGA CTG GAG AAA   bp206
 K   I   R   H   R   K   T   T   R   R   G   E   M   E   T   L   E   P   N   N   L   W   Y   L   F   T   G   L   E   K   aa688

GGA AGT CAG TAC AGT TTC CAG GTG TCA GCC ATG ACA GTC AAT GGT ACT GGA CCA CCT TCC AAC TGG TAT ACT GCA GAG ACT CCA GAG AAT   bp215
 G   S   Q   Y   S   F   Q   V   S   A   M   T   V   N   G   T   G   P   P   S   N   W   Y   T   A   E   T   P   E   N   aa718

GAT CTA GAT GAA TCT CAA GTT CCT GAT CAA CCA AGC TCT CTT CAT GTG AGG CCC CAG ACT AAC TGC ATC ATC ATG AGT TGG ACT CCT CCC   bp224
 D   L   D   E   S   Q   V   P   D   Q   P   S   S   L   H   V   R   P   Q   T   N   C   I   I   M   S   W   T   P   P   aa748

TTG AAC   bp225
 L   N     aa750
```

Figure 12-2. The nucleotide sequence and the predicted amino acid sequence of its "open reading frame" (a sequence that specifies a corresponding sequence of amino acids) found in DNA cloned from a human tissue culture cell line. Sixteen overlapping fragments were analyzed and the results were then assembled to yield this 2250-base-pair sequence. At least a dozen scientific journals (and several computer data banks) are devoted to reporting and storing data of this sort. Open reading frames such as this one are probably structural genes whose function may or may not yet be known. (bp = base pair; aa = amino acid.)

artifacts by matching the growth rings of trees. Although an individual tree may live no more than 200–300 years, by matching the growth rings of living trees with trees long dead but not decayed (desert areas are the usual sites for such studies) or with lumber used by primitive (even prehistoric) peoples in constructing houses, sleds, boats, and other items, archaeologists have developed records that span thousands of years. The bristle-cone pine, an extremely long-lived tree of the western United States, has provided enough information for dendrochronologists to create a record going back nearly 10,000 years. If tree rings can permit the construction of such a record, matching sequences of base pairs in DNA fragments are likely to lead to an unraveling of the base pair sequences in "terribly long" segments of DNA.

The sequence of DNA represented in Figure 12-2 is 2250 base pairs long. There are an estimated 4 billion base pairs in the human genome. Consequently, nearly 1.8 million figures as complex as that shown in Figure 12-2 would be needed to illustrate the entire genome. At 425 pages per volume (1 figure on each page, no text), 4200 volumes would be required to tell the whole story. The termination of that achievement—during the year 2005, according to a quotation I cited in Chapter 1—will represent the end of the search for the gene in the sense that I have viewed that search. All the genetic information pertaining to human beings will be available when that enormous set of volumes (with numerous footnotes in which individual variants are noted) has been completed.

I visualize the human genome as the torn pages of a giant novel, written in an unknown language, blowing about helter-skelter in an air-conditioned, enclosed space such as Houston's Astrodome. The pages blow about, but they do not blow away, nor do they rot. The Human Genome Project resembles an effort to retrieve these thousands of blowing, fluttering scraps of paper and to arrange them into whole pages and sequences of pages. The job is not an impossible one. Recall that the radical fundamentalist students in Iran did precisely that with the shredded records they found in the American embassy in Teheran. The outcome is a set of volumes that sells extremely well in bookstores throughout the Middle East.

The Iranians, however, were familiar with English. Those working on the genome project know the rules (discovering fragmented sequences and joining different sequences into a single larger one) by which the grand sequence that extends from the first base pair of chromosome 1 to the last base pair of chromosome 23 will eventually emerge. That corresponds to assembling the novel. The language is partially known: the rules by which DNA specifies amino acid sequences are known. Figure 12-2 provides an excellent example. The codon for methionine is the start signal for protein synthesis; the amino acid may be left on the finished protein or it may be snipped off by an enzyme. Following that starting codon, 749 subsequent ones specify additional amino acids. A sequence that long is unlikely to occur unless one is actually in an open reading frame; that is, unless one is really examining DNA that constitutes a structural gene.

Three of the 64 possible triplets of bases are stop codons. If DNA is not read properly, the average number of codons encountered that specify an amino acid before encountering a termination codon by chance is about 20 (i.e., 64 ÷ 3). To check this claim, I will take the line in Figure 12-2 labeled "bp 1704" and, omitting its opening *A,* write down the remaining bases in sets of three:

CCT CAC CTA CCT CAA TTC TTA TTA CCT GGG

The mRNA transcribed from this DNA would be:

GGA GUG GAU GGA GUU AAG AAU AAU GGA CCC

The sequence of amino acids would be:

Gly-Val-Asp-Gly-Val-Lys-Asn-Asn-Gly-Pro

This sequence differs at every position from the one listed in the figure for the same region read properly:

Thr-Ser-Pro-Thr-Ser-Ile-Leu-Ile-Thr-Try

But a stop codon was not encountered: the odds are actually better than fifty-fifty that I should not have encountered 1 in 10 randomly generated codons. Recalling, however, that the DNA sequences that specify the stop codons UGA, UAA, and UAG are, respectively, ACT, ATT, and ATC, we can identify several potential stop codons in the line labeled "bp 354" alone. Had the reading frame been shifted by either one or two bases, right or left, a long sequence of "meaningful" codons would not have been encountered.

Although molecular biologists are adept at translating (as a mental exercise) base pair sequences into amino acid sequences, they are still struggling to obtain a complete understanding of upstream controls—especially in higher organisms. Certain consistent patterns have emerged from their observations, to be sure:

1. A sequence often encountered in the upstream control region is TATA the "TATA" box. This sequence seems to be involved in positioning the enzyme RNA polymerase so that it starts at the right place when it passes down the DNA and transcribes messenger RNA.

2. A second sequence of bases frequently encountered in the control region of DNA is CAAT, the "CAAT" box. This sequence can be oriented as either CAAT or TAAC and seems to be involved with the efficiency of the promoter region; that is, with the number of RNA polymerase molecules that attach to the region during a given length of time.

3. A third commonly encountered sequence is GGGCGG, the "GC" box. This sequence can occur several times in a promoter region and can be oriented in either direction.

4. Various *enhancers* of transcription can be found near the gene—upstream (at times), within the gene, or downstream from the gene. As the name implies, the rate of transcription is enhanced, regardless of the element's actual position. How the enhancer performs its task is unknown.

Figure 12-3 illustrates how single base pair substitutions within DNA control regions are used to probe the role these regions play in regulating gene activity. In addition to the rates of transcription and the accuracy of transcription that I emphasized in the above catalog, control also involves the tissues or cells within which the activity of a gene is expressed, if it is expressed at all. Understanding all of the details of gene control will be extremely difficult. Recall the 20 or more major bristles on a Drosophila fly—40 or more if both right and left sides are considered. Each of these has a *precise* position on a normal fly. Each arises from a single cell chosen in some unknown manner from the thousands of cells that form the epidermis of a fly. Each completes one-third or more of its task of forming a gigantic (for it) bristle within a 24-hour period. Here is but one small example of the sophistication of gene control. Problems involving sheer bulk, like the synthesis of hemoglobin in a man, an elephant, or a whale, are not the difficult ones to solve; the difficult ones are those that reflect a precision or a delicacy that is hard even to describe.

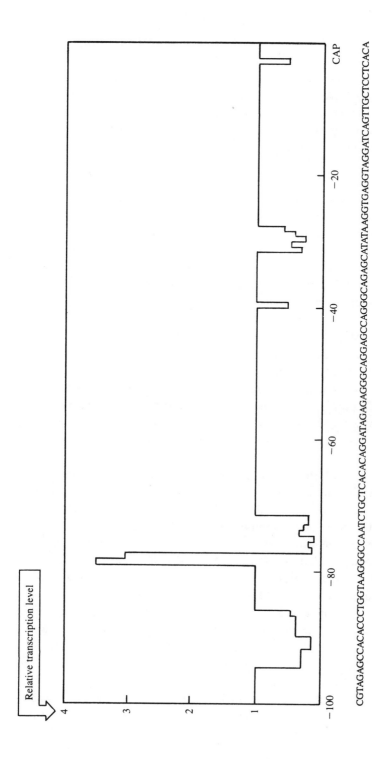

Figure 12-3. A base-by-base analysis of the effect of mutational substitutions in the upstream control region on the rate or efficiency of transcription of an adjacent structural gene (β-globin). Of special interest are the TATA, CAAT, and GC "boxes" discussed in the text. (Redrawn from Lewin, 1990, by permission of Benjamin Lewin.)

CGTAGAGCCACACCCTGGTAAGGGCCAATCTGCTCACACAGGATAGAGAGGGCAGGAGCCAGGGCAGAGCATATAAGGTGAGGTAGGATCAGTTGCTCCTCACA

The Human Genome Project: A personal view

Earlier, I likened the human genome to the torn pages of an unknown novel—a novel written in an utterly foreign language—strewn about within the Houston Astrodome but with no danger of losing even one page. I have described (Box 12-1) the procedures by which DNA can be sequenced, techniques that are as simple and as basic as those required for performing genetic crosses of either *Drosophila, Zea mays,* or *Neurospora.* Hence, there is nothing that prohibits the undertaking of the Human Genome Project except doubts as to whether it is worthwhile. The entire genome, as I pointed out, will require 1.8 million figures similar to that reproduced as Figure 12-2; 4200 volumes of 425 pages each—each looking much like Figure 12-2; and each one acquired not by the application of imaginative thought but by technician-scientists performing routine chores no more complicated than mating flies, pollinating corn, or culturing yeast and molds.

Financial resources are not unlimited. I know of one zoological park that recently lost 10% of its budget so that another county agency could provide care for drug-addicted newborn babies. Congress argues over whether foreign aid should be diverted from old friends (and, by now, dependents) to the newly liberated countries of Eastern Europe. The pot of funds available for any collection of enterprises is not unlimited.

Why must the human genome be completely known by 2005? Not because it will otherwise be lost! If it is lost, human beings are also lost because, in this case, *we are the novel.* If human beings continue to exist on earth, the human genome will continue as well. If we cease to exist, who really cares about our genomes? The only reason the Human Genome Project has been assigned a high priority is that certain persons alive today want to see the entire sequence of base pairs for themselves and perhaps have a historical advantage in perceiving the meaning of certain sequences of these base pairs. The urgency involves personal claims of prior discovery.

The Azores, a chain of nine islands lying in the mid-Atlantic Ocean, were discovered by the Portuguese on August 15, 1432. That is, Santa Maria was discovered at that time. History books record that the other islands were discovered years later, although on clear days one of them, São Miguel, is visible from Santa Maria. The point is that today, at the close of the twentieth century, no one cares whether São Miguel was reached within days or years after the first Portuguese ships arrived at Santa Maria.

There sat São Miguel—visible, even if not clearly so—and it was not about to disappear. The same applies to the human genome. Years from now no one will either remember or care whether the base sequence of human DNA was determined during or after the twentieth century.

The cost of the Human Genome Project has been estimated at $3 billion. The bulk of this money will be spent for the salaries of persons who will do nothing more than perform repetitive tasks that, although they involve molecular techniques, have been compared intellectually with the genetic crosses—the drudgery—of yesteryear. How else might this money be spent? At $30,000 per grant, 100,000 young scientists could be supported during the same interval for the same amount of money. More accurately, I should say that 33,000 young scientists could be awarded three-year grants of approximately $30,000 per year. Intellectual advances are made by young investigators who are encouraged to ask their own questions and to attempt to answer them by their own means. Advances in science come from inquisitive minds struggling to unravel intriguing secrets of nature. In my estimation, science, science education, and the public understanding of scientific research would be advanced much more by the support of small, personal, and imaginative research projects than by the investment of huge sums in a massive but intellectually uninspired enterprise such as the Human Genome Project. In short, although the Human Genome Project was the snow-capped peak that was sighted early in Chapter 1 and that subsequently guided our search for the gene, I see no compelling need for attempting the final assault on that peak today. Instead, let's enlarge our latest base camp and make it permanent. Let's allow both young and old to become familiar with the nearby terrain. Let's permit the climbers among the young scientists to practice on neighboring slopes. Let's *explore*. The major peak will be scaled eventually, but the costs will be lessened, and the rewards increased, if the assault is a deliberate and unhurried one.

Box 12-1. Sequencing DNA

In Box 2-3, I made an effort to describe—at least superficially—the basis of light microscopy. The same effort was not attempted for electron microscopy because magnets and electron beams are not part of everyday life; hand lenses, cameras, projectors, and eyeglasses are much more familiar. I have, however, pointed out that the sequencing of DNA has been accomplished by biochemical (molecular) procedures, not by the invention of a giant microscope. These biochemical procedures are not impossible to grasp, and, because of their importance in the study of DNA, a commonly used one is described here.

To start, one might isolate a segment of DNA and insert it into *E. coli* as part of an artificially constructed single-stranded DNA. This may sound mysterious and difficult, but technical manipulations always seem complex to those who don't use them routinely.

With the single-stranded DNA of *one* sort available, one can arrange to have the complement of that strand synthesized over and over again until large amounts of only the complement are available. Indeed, one can carry out identical syntheses in four separate tubes. The next step involves a chemical trick. The complementary strand of DNA is constructed from its model; let's say, from left to right. First a thymine is inserted, then another, then an adenine, and so on—each base being specified by the corresponding one on the model, each one being attached to the preceding one of the growing strand by virtue of a particular oxygen atom in the pentose sugar of the one and the phosphoric acid of the newcomer (see Figures 6-7 and 8-4). The trick used by biochemists is to add to the chemical mixture a terminator substance: *dideoxyribonucleic acid*. The prefix *di-* means that a second atom of oxygen has been removed from the pentose sugar; the second missing oxygen is the one that would otherwise react with the phosphoric acid of the incoming base. Without this oxygen atom, the growing strand of DNA can no longer elongate; it terminates (see Figure 6-7).

Four terminator dideoxy's are available: (ddATP), (ddTTP), (ddGTP), and (ddCTP). (Note that A, T, G, and C refer to adenine, thymine, guanine, and cytosine; TP in each case refers to triphosphate, the energy source.) When added in low concentration, one to each of the four tubes, there is a correspondingly low probability that a terminator molecule will be added to the growing DNA strand at any appropriate position. Once inserted at that position, the strand no longer grows; it aborts instead.

The DNA that is being copied may be written as follows:

AATGACTCCGTAAGGCG

Copies of this strand may be terminated by the introduction of ddATP at any site where a T occurs (A* represents the terminating adenine):

	Number of bases in fragment
TTA*	3
TTAGTCA*	7
TTAGTCAGGCA*	11

In a second tube, ddCTP is used as the strand terminator; again the strand stops growing wherever the terminating cytosine (C*) is inserted.

TTAC*	4
TTACTGAGGC*	10
TTACTGAGGCATTC*	14
TTACTGAGGCATTCC*	15
TTACTGAGGCATTCCGC*	17

In a third tube, ddGTP is the terminator; it is inserted wherever the model contains C:

TTACTG*	6
TTACTGAG*	8
TTAGTGAGG*	9
TTACTGAGGCATTCCG*	16

In the fourth tube, ddTTP is used:

T*	1
TT*	2
TTACT*	5
TTACTGAGGCAT*	12
TTACTGAGGCATT*	13

Notice what has been generated: 17 fragments, a number corresponding to the 17 purines and pyrimidines in the single-stranded primer. Furthermore, the 17 fragments are of 17 different lengths ranging from 1 base to all 17. High-resolution electrophoresis can clearly separate fragments that differ in length by a single base.

When the newly synthesized contents of each of the four tubes are electrophoresed in adjacent channels one gets the following results, which correspond with those illustrated in Figure 12-1:

	A	C	G	T
17		—		
16			—	
15		—		
14		—		
13				—
12				—
11	—			
10		—		
9			—	
8			—	
7	—			
6			—	
5				—
4		—		
3	—			
2				—
1				—

Reading these channels from the bottom up, one obtains TTACTGAG-GCATTCCGC, which proves to be the complementary strand of the single-stranded DNA listed at the start of this exercise. Earlier, I said one could read the sequence by dragging a finger up the gel. I am old-fashioned! One can buy an automatic scanner that reads gels of this sort just as the scanner at the checkout counter in the supermarket reads the bar code on each purchase. Modern biology is increasingly an automated science.

13 Epilogue

Cub reporters on their first assignments are reminded by their editors to cover the following aspects of each story: What? Where? When? Why? Who? and How? Describing the search for the gene has not been unlike a news assignment:

What is the physical nature of the hypothetical factors whose pattern of inheritance was studied and eventually unraveled first by Gregor Mendel and subsequently by a host of other persons known as geneticists?

Where are these factors located? Of what are they composed? The search led us to the Morgan school and the chromosome theory, through microbial genetics, and into phage genetics and molecular biology.

When did the search for the gene begin? It started with (or even before) the origins of agriculture and has proceeded at an ever-increasing pace until today. If we can believe the authors of the textbook *Molecular Biology of the Gene* (Watson et al., 1987), events are now occurring at a pace such that priorities will soon be established by the stopwatch rather than by the calendar as in the past.

Why have so many persons for so many years searched for the gene? I prefer to believe that the search for the gene was an attempt to understand nature, an attempt to solve an otherwise baffling array of everyday observations. I am not unaware, however, of the additional view: knowledge is power! To understand means in large measure to control. Agriculturalists, in their attempts to improve domesticated plants and animals, have practiced artificial selection for millennia. Knowledge concerning patterns and the physical basis of inheritance makes the practice of selection much more efficient. Eugenicists, the would-be modifiers of human populations, also view genetic information as a source of power: the power to eliminate unwanted genetic traits; the power to increase the proportions of wanted ones. Unfortunately, "power" in the practice of eugenics has often meant political power as well as scientific knowledge, hence the unpleasant aura surrounding much of eugenics and many of its practitioners. Finally, medical doctors eagerly await advances in molecular genetics in order to gain power over human disease.

Who did it all? In recounting the search for the gene, I have cited numerous—but still a finite number of—names. These names are in the index and among the References and Selected Readings at the end of this book. They are in many other textbooks as well. No one should imagine, however, that these are the only people who succeeded in their quest for truth, nor that they are in some manner superior beings. Each person who contributed to our understanding of the gene was instructed by his or her predecessors and was stimulated by conversations and discussions with friends and colleagues. Each stood on the shoulders of teachers and colleagues. In science, as in the natural world, there are no Adams and Eves who arise by acts of special creation. Science, including genetics, is an undertaking that relies on the exchange and recombination of ideas held by many people of many cultures. Such an exchange of ideas is to cultural heredity what sexual reproduction is to physical heredity. Great ideas do not arise in isolation; tracing the origins of great ideas, like reconstructing one's family pedigree, is a nearly hopeless task.

How does the gene function? (I have already answered the parallel question: How have geneticists of successive eras conducted their search for the gene?) Studies on gene action have paralleled the search for the gene itself since the rediscovery of Mendel's 1865 paper. By modern standards the early investigations of gene action were necessarily crude; first, because the concept of the gene itself was crude, and, second, because technical (largely chemical) procedures were crude. The following quotation illustrates these points: "The isolation of the first repressors and the demonstration that they bind specifically to control sequences in DNA seemed to some pioneers in DNA research to mark the end of the years of germinal discovery. With no means to isolate the genes of any higher organism, much less any way to know their nucleotide sequences, any pathway to understanding *how* genes guide the differentiation events that give rise to multicellular organisms seemed impossibly remote" (Watson et al., 1987, p. v; emphasis added).

The *how* of the above quotation is the tail that now wags the dog! This can be illustrated no more clearly than by citing a second quotation from the same reference and page as the first:

> It is only in this fourth edition [two volumes, 1163 oversized pages] that we see the extraordinary fruits of the recombinant DNA revolution. Hardly any contemporary experiment on gene *structure* or *function* is done today without recourse to ever more powerful methods for cloning and sequencing genes. As a result, we are barraged daily by arresting new facts of such importance that we seldom can relax long enough to take comfort in the accomplishments of the immediate past. The science described in this edition is by any measure an extraordinary example of human achievement. (Emphasis added)

The emphasis placed on gene *structure* and *function* in the preceding quotation underscores the fact that the search for the gene has ended. The gene is now being dissected (a process that reveals its structure) in order to determine how it interacts with RNA, proteins, and other molecules in carrying out its various functions. *How* a gene functions is a matter that is intellectually separate from the earlier questions What? and Where? Now that we know what the gene is, of what is it composed, and the manner in which it stores information concerning both the structure of proteins and the timing of their synthesis, the study of gene action becomes an exercise in chemistry—albeit, the chemistry of extremely complex organic molecules and their interactions.

References

References that have exerted a general influence over this text beyond the citation of one or more specific items are marked by an asterisk; in this manner, I hope to convey my thanks and appreciation.

Andrewartha, H. G. 1963. *Introduction to the Study of Animal Populations.* Chicago: University of Chicago Press.

Anfinsen, C. B. 1959. *The Molecular Basis of Evolution.* New York: John Wiley and Sons.

Avery, O. T., C. M. MacLeod, and M. McCarthy. 1944. Studies on the chemical nature of the substance inducing transformation of pneumococcal types. J. Exp. Med. 79:137–157.

Bateson, W. 1916. *The Mechanism of Mendelian Heredity,* by T. H. Morgan, A. H. Sturtevant, H. J. Muller, and C. B. Bridges (a review). Science 44:536–543.

Benzer, Seymour. 1955. Fine structure of a genetic region in bacteriophage. Proc. Natl. Acad. Sci. USA 41:344–354.

Benzer, Seymour. 1957. The elementary units of heredity. In W. D. McElroy and Bentley Glass, eds., *The Chemical Basis of Heredity,* pp. 70–93. Baltimore: Johns Hopkins University Press.

Boveri, Th. 1902. Über mehrpolige Mitosen als Mittel zur Analyse des Zellkerns. Verh. Physikal. Med. Ges. Würzburg 35:67–90.

Brady, J. P., and R. C. Richmond. 1990. Molecular analysis of evolutionary changes in the expression of *Drosophila* esterases. Proc. Natl. Acad. Sci. USA 87:8217–8221.

Bridges, C. B. 1916. Non-disjunction as proof of the chromosome theory of heredity. Genetics 1:1–52, 107–163.

Bridges, C. B. 1935. Salivary chromosome maps. With a key to the banding of the chromosomes of *Drosophila melanogaster.* J. Hered. 25:60–64.

Bridges, C. B., E. N. Skoog, and J. C. Li. 1936. Genetical and cytological studies of a deficiency (*Notopleural*) in the second chromosome of *Drosophila melanogaster*. Genetics 21:788–795.

Brink, R. A., and D. C. Cooper. 1935. A proof that crossing over involves an exchange of segments between homologous chromosomes. Genetics 20:22–35.

Britten, R. J., and E. H. Davidson. 1969. Gene regulation for higher cells: A theory. Science 165:349–357.

Cairns, John. 1963. The chromosome of *Escherichia coli*. Cold Spring Harbor Symp. Quant. Biol. 28:43–46.

Castle, W. E. 1919. Is the arrangement of the genes in the chromosome linear? Proc. Natl. Acad. Sci. USA 5:25–32.

Chargaff, Erwin. 1950. Chemical specificity of nucleic acids and mechanism of their enzymatic degradation. Experientia 6:201–209.

Creighton, H. B., and B. McClintock. 1931. A correlation of cytological and genetical crossing-over in *Zea mays*. Proc. Natl. Acad. Sci. USA 17:492–497.

Creighton, H. B., and B. McClintock. 1935. The correlation of cytological and genetical crossing-over in *Zea mays*. A corroboration. Proc. Natl. Acad. Sci. USA 21:148–150.

Crick, F. H. C. 1962. The genetic code. Sci. Am. 207:66–74.

Crick, F. H. C. 1966. The genetic code: Yesterday, today and tomorrow. Cold Spring Harbor Symp. Quant. Biol. 31:3–9.

Darlington, C. D., and K. Mather. 1949. *The Elements of Genetics*. New York: Macmillan.

Darwin, Charles. 1875. *The Variation of Animals and Plants under Domestication*, 2d ed. John Murray: London.

Davey, R. B., and D. C. Reanney. 1980. Extrachromosomal genetic elements and the adaptive evolution of bacteria. Evol. Biol. 13:113–147.

Dobzhansky, Th., and Ernst Boesiger. 1983. *Human Culture: A Moment in Evolution*. New York: Columbia University Press.

Edgar, R. S., and R. H. Epstein. 1965. *The Genetics of a Bacterial Virus*. New York: Scientific American.

Fearen, E. R., K. R. Cho, J. M. Nigro, and others. 1990. Identification of a chromosome 18q gene that is altered in colorectal cancers. Science 247:49–56.

Flemming, Walther. 1882. *Zellsubstanz, Kern und Zelltheilung*. Leipzig: Verlag von F. C. W. Vogel.

Frisch, Karl von. 1950. *Bees: Their Vision, Chemical Senses and Language*. Ithaca, N.Y.: Cornell University Press.

Glass, H. Bentley. 1947. Maupertuis and the beginnings of genetics. Q. Rev. Biol. 22:196–210.

Grene, Marjorie, and R. M. Burian. 1991. Wilhelm Roux: early historical inferences regarding the "Why?" of nuclear division patterns, with a translation of Roux (1883). Evol. Biol. 25:427–444.

Greslin, A. F., D. M. Prescott, Y. Oka, S. H. Loukin, and J. C. Chappell. 1989. Reordering of nine exons is necessary to form a functional actin gene in *Oxytricha nova*. Proc. Natl. Acad. Sci. USA 86:6264–6268.

Haldane, J. B. S. 1939. *The Marxist Philosophy and the Sciences.* New York: Random House.

Haldane, J. B. S. 1964. A defense of beanbag genetics. Perspect. Biol. Med. 7:343–359.

Harlan, Jack R. 1975. *Crops and Man.* Madison, Wis.: Am. Soc. Agron./Crop Sci. Soc. Am.

Henking, H. 1891. Untersuchungen über die ersten Entwicklungsvorgänge in den Eiern der Insekten. Z. Wiss. Zool. 51:685–736.

Hershey, A. D., and M. Chase. 1952. Independent function of viral protein and nucleic acid in growth of bacteriophage. J. Gen. Physiol. 36:39–56.

Hertwig, O. 1890. Vergleich der Ei- und Samenbildung bei Nematoden. Eine Grundlage fur zelluläre Streitfragen. Arch. Mikrosk. Anat. 36:1–138.

Hilu, K. W., and J. M. J. deWet. 1980. Effect of artificial selection on grain dormancy in *Eleusine* (Gramineae). Syst. Bot. 5:54–60.

Hirsh, J. 1989. Molecular genetics of dopa decarboxylase and biogenic amines in *Drosophila.* Dev. Genet. 10:232–238.

Ingram, V. M. 1963. *The Hemoglobins in Genetics and Evolution.* New York: Columbia University Press.

Kaufmann, B. P., and R. C. Bate. 1938. An X-ray induced intercalary duplication in *Drosophila* involving the union of sister chromatids. Proc. Natl. Acad. Sci. USA 24:368–371.

Knapp, E., and H. Schreiber. 1939. Mutations induced in *Sphaerocarpus* by ultraviolet light. Proc. 7th Int. Congr. Genet., p. 175.

Laxton, Thomas. 1872. Notes on some changes and variations in the offspring of cross-fertilized peas. J. R. Hort. Soc. Lond. 3:10–14.

Lewin, Benjamin. 1990. *Genes IV.* New York: Oxford University Press.

Lindsley, D. L., and E. H. Grell. 1968. *Genetic Variations of* Drosophila melanogaster. Carnegie Inst. Wash. Publ. 627:1–471.

Mayr, Ernst. 1963. *Animal Species and Evolution.* Cambridge, Mass.: Harvard University Press.

McClintock, Barbara. 1965. The control of gene action in maize. Brookhaven Symp. Biol. 18:162–184.

McClintock, Barbara. 1967. Genetic systems regulating gene expression during development. Dev. Biol. Suppl. 1:84–112.

Mendel, Gregor. 1865. Experiments in plant hybridization. Verh. Naturforsch. Ver. Brunn 4 (Abhandlungen):3–47. Translated by the Royal Horticultural Society of London.

Menozzi, P., A. Piazza, and L. Cavalli-Sforza. 1978. Synthetic maps of human gene frequencies in Europeans. Science 201:786–792.

Meselson, M., and F. W. Stahl. 1958. The replication of DNA in *Escherichia coli.* Proc. Natl. Acad. Sci. USA 44:671–682.

Mirsky, A. E. 1951. Some chemical aspects of the cell nucleus. In L. C. Dunn, ed. *Genetics in the 20th Century,* pp. 127–153. New York: Macmillan.

*Moore, J. A. 1963. *Heredity and Development.* New York: Oxford University Press.

Morgan, T. H. 1909. What are 'factors' in Mendelian explanations? Proc. Am. Breeders' Assoc. 5:365–368.

Morgan, T. H. 1940. Calvin Blackman Bridges. Genetics 25:i–v.

Morgan, T. H. 1942. Genesis of the white-eyed mutant. J. Hered. 33:91–92.

Muller, H. J. 1929. The gene as the basis of life. Proc. Int. Congr. Plant Sci. 1:897–921.

Muller, H. J. 1961. Genetic nucleic acid: Key material in the origin of life. Perspect. Biol. Med. 5:1–23.

Muller, H. J. 1962. *Studies in Genetics.* Bloomington: Indiana University Press.

Nirenberg, M. W., and J. H. Matthaei. 1961. The dependence of cell-free protein synthesis in *E. coli* upon naturally occurring or synthetic polyribonucleotides. Proc. Natl. Acad. Sci. USA 47:1588–1602.

Olby, Robert. 1985. *Origins of Mendelism,* 2d ed. Chicago: Chicago University Press.

Patau, K. 1935. Chromosomenmorphologie bei *Drosophila melanogaster* und *Drosophila simulans* und ihre genetische Bedeutung. Naturwissenschaften 23:537–543.

Pauling, Linus, H. A. Itano, S. J. Singer, and I. C. Wells. 1949. Sickle cell anemia: a molecular disease. Science 110:543–548.

*Portugal, F. H., and J. S. Cohen. 1977. *A Century of DNA.* Cambridge, Mass.: MIT Press.

Punnett, R. C. 1928. *Scientific Papers of William Bateson,* vol. 2. Cambridge: Cambridge University Press. Reprint. New York: Johnson Reprint, 1971.

Roberts, L. 1990. Research news: The Worm Project. Science 248:1310–1313.

Ruiz-Gomez, Mar, and Juan Modolell. 1987. Deletion analysis of the *achaete-scute* locus of *Drosophila melanogaster.* Genes Dev. 1:1238–1246.

Schneider, A. 1873. Untersuchungen über Plathelminthen. Oberhess. Gesell. Natur. Heilkd. 14:69–140.

Sena, E. P. 1966. Developmental variation of alkaline phosphatases in *Drosophila melanogaster.* Master's thesis, Cornell University, Ithaca, N.Y.

Sherwood, E. R. 1966. Gregor Mendel: Experiments on plant hybrids. In Curt Stern and E. R. Sherwood, eds., *The Origin of Genetics: A Mendel Source Book,* pp. 113–149. San Francisco: W. H. Freeman.

Sinnott, E. W., L. C. Dunn, and Th. Dobzhansky. 1958. *Principles of Genetics.* New York: McGraw-Hill.

Slizynska, H. 1938. Salivary gland analysis of the white-facet region of *Drosophila melanogaster.* Genetics 23:291–299.

Srb, Adrian, and R. D. Owen. 1952. *General Genetics.* San Francisco: W. H. Freeman.

Stadler, L. J., and F. Uber. 1942. Comparison of genetic effects of different wavelengths of ultra-violet light on maize. Genetics 27:84–118.

Stahl, F. W. 1964. *The Mechanics of Inheritance.* Englewood Cliffs, N.J.: Prentice-Hall.

Stemler, A. B. L., J. R. Harlan, and J. M. J. deWet. 1975. Caudatum sorghums and the speakers of Chari-Nile languages in Africa. J. Afr. Hist. 16:161–183.

Stent, G. S. 1963. *Molecular Biology of Bacterial Viruses.* San Francisco: W. H. Freeman.

Stent, Gunther, ed. 1980. *The Double Helix,* by James D. Watson (with editor's comments and reviews). New York: W. W. Norton.

*Stubbe, Hans. 1972. *History of Genetics.* Cambridge, Mass.: MIT Press.

*Sturtevant, A. H. 1965. *A History of Genetics.* New York: Harper and Row.

Sutton, W. S. 1903. The chromosomes in heredity. Biol. Bull. 4:231–251.

Wallace, B. 1966. *Chromosomes, Giant Molecules, and Evolution.* New York: W. W. Norton.

Wallace, B. 1984. Changes in the genetic mentality. Perspect. Biol. Med. 27:598–610.

Wallace, B. 1988. In defense of verbal arguments. Perspect. Biol. Med. 31:201–211.

Wallace, B., and G. M. Simmons, Jr. 1987. *Biology for Living.* Baltimore: Johns Hopkins University Press.

*Watson, J. D. 1965. *Molecular Biology of the Gene.* New York: W. A. Benjamin.

Watson, J. D. 1968. *The Double Helix.* New York: New American Library. G. S. Stent (1980) has edited a special edition of *The Double Helix* in which are included the original text, commentaries, reviews, and original papers. New York: W. W. Norton.

Watson, J. D., and F. H. C. Crick. 1953a. A structure for desoxyribose nucleic acids. Nature (Lond.) 171:737–738.

Watson, J. D., and F. H. C. Crick. 1953b. The structure of DNA. Cold Spring Harbor Symp. Quant. Biol. 18:123–131.

Watson, J. D., N. H. Hopkins, J. W. Roberts, J. A. Steitz, and A. M. Weiner. 1987. *Molecular Biology of the Gene,* 4th ed. New York: Benjamin-Cummings.

White, M. J. D. 1948. *Animal Cytology and Evolution.* Cambridge: Cambridge University Press.

Whiting, P. W. 1935. Sex determination in bees and wasps. J. Hered. 26:263–278.

Wilson, E. B. 1895. *An Atlas of the Fertilization and Karyokinesis of the Ovum.* New York: Columbia University Press.

Wilson, E. B. 1918. Theodor Boveri. In *Erinnerungen an Theodor Boveri.* Ed. W. C. Röntgen. Tubingen: J. C. B. Mohr. [Dorothea Rudnick, trans. 1967. Theodor Boveri. Berkeley: University of California Press.]

Woese, Carl R. 1984. *The Origin of Life.* Carolina Biology Readers 13, ed. J. S. Head. Burlington, N.C.: Carolina Biological Supply.

Wright, T. R. F., and R. MacIntyre. 1963. A homologous gene-enzyme system, esterase-6, in *Drosophila melanogaster* and *D. simulans.* Genetics 48:1717–1726.

Yanofsky, C., B. C. Carlton, J. R. Guest, D. R. Helinski, and U. Henning. 1964. On the colinearity of gene structure and protein structure. Proc. Natl. Acad. Sci. USA 51:266–272.

Yanofsky, S. A., and Sol Spiegelman. 1963. Distinct cistrons for the two ribosomal RNA components. Proc. Natl. Acad. Sci. USA 49:538–544.

Selected Readings

The readings listed here complement the publications that I found to be especially useful and that are marked with asterisks in the References. The purpose to which these readings are put depends upon the reader. If *The Search for the Gene* has been assigned to biology students to provide a historical background for the thousand or so pages of technical information contained in standard genetics textbooks, the selected readings merely provide guidance to the intellectually curious among them. For "non-majors" using *The Search for the Gene* as a textbook, these readings provide an opportunity to view this search through the eyes of other observers: early geneticists, other biologists who played peripheral roles, members of the Phage Group, and professional historians. Among the suggested readings are some in which the authors place modern genetics into broader-than-usual contexts—both personal contexts and those of modern society.

Beadle, G. W. 1963. *Genetics and Modern Biology.* Philadelphia: American Philosophical Society.

Cairns, J., G. S. Stent, and J. D. Watson. 1966. *Phage and the Origins of Molecular Biology.* Cold Spring Harbor, N.Y.: C. S. H. Laboratory of Quantitative Biology. The essay by J. D. Watson in this volume is a historical introduction to his later book, *The Double Helix.* A sense of the intensity and caliber of the phage research can be obtained from the essay by N. ("Nick") Visconti, younger brother of the movie producer and an excellent scientist.

Carlson, E. A. 1966. *The Gene: A Critical History.* Philadelphia: Saunders. The author is a geneticist, a student of H. J. Muller, and a noted university teacher.

Carlson, E. A. 1972. H. J. Muller. Genetics 70:1–30. The journal *Genetics* has for many years featured the obituary of an outstanding geneticist in the opening pages of each volume.

Chargaff, Erwin. 1963. *Essays on Nucleic Acids*. Amsterdam: Elsevier. An advanced but urbane, charming, and thoughtful collection of essays.

Crick, Francis. 1970. *Of Molecules and Men*. Seattle: University of Washington Press.

Crick, Francis. 1981. *Life Itself: Its Origin and Nature*. New York: Simon and Schuster.

Crick, Francis. 1988. *What Mad Pursuit: A Personal View of Scientific Discovery*. New York: Basic Books. A personal recollection of his role in the discovery of the double helix.

deBusk, A. G. 1968. *Molecular Genetics*. New York: Macmillan.

Delbrück, Max. 1949. A physicist looks at biology. Trans. Conn. Acad. Arts Sci. 38:173–190. A pre–double helix view of biology.

Delbrück, Max. 1971. Aristotle-totle-totle. In J. Monod and E. Borek, eds., *Of Microbes and Life*. New York: Columbia University Press. An essay not about the search for the gene but one of many wide-ranging essays dedicated to André Lwoff, a prominent leader in that search.

Delbrück, Max. 1986. *Mind from Matter? An Essay on Evolutionary Epistemology*. Cambridge, Mass.: Blackwell Scientific. Lectures to an elementary class that include DNA and molecular genetics as givens.

Dubos, René. 1976. *The Professor, the Institute, and DNA*. New York: Rockefeller University Press. About Oswald T. Avery and his role in the discovery that DNA is the transforming principle.

Dunn, L. C. 1965. *A Short History of Genetics*. New York: McGraw-Hill.

Dunn, L. C., ed. 1951. *Genetics in the Twentieth Century*. New York: Macmillan.

Frankel, Edward. 1979. *DNA, the Ladder of Life*. New York: McGraw-Hill.

Freifelder, David. 1978. *The DNA Molecule*. San Francisco: W. H. Freeman. A collection of nearly fifty papers, many of which have been cited in the present book, and each of which is accompanied by the author's analysis and relevant questions.

Goldschmidt, R. B. 1956. *Portraits from Memory: Recollections of a Zoologist*. Seattle: University of Washington Press. Recollections of an iconoclastic geneticist that touch on Boveri and other non-American pioneers.

Gonick, Larry, and Mark Wheelis. 1983. *The Cartoon Guide to Genetics*. New York: Harper and Row. Formal genetics in a different light.

Gribben, John. 1985. *In Search of the Double Helix: Quantum Physics and Life*. New York: McGraw-Hill. Both Darwin and quantum physics are made parts of the search.

Hartman, P. E., and S. R. Suskind. 1969. *Gene Action*. Englewood Cliffs, N.J.: Prentice-Hall. An early text on a subject that this volume has largely ignored.

Hoagland, M. B. 1978. *The Roots of Life: A Layman's Guide to Genes, Evolution, and the Ways of Cells*. Boston: Houghton Mifflin.

Hoagland, M. B. 1981. *Discovery, the Search for DNA's Secrets*. Boston: Houghton Mifflin.

Hunt, Tim, Steve Prentis, and John Tooze. 1983. *DNA Makes RNA Makes Protein*. Amsterdam: Elsevier Biomedical Press. Somewhat advanced essays but pertinent to the search.

Jacob, François. 1973. *The Logic of Life*. New York: Pantheon Books. An eloquent review of genetics as an integral fact of life.

Jacob, François. 1982. *The Possible and the Actual*. Seattle: University of Washington Press. Includes a chapter dealing with evolution as a tinkerer—not an engineer.

Judson, H. F. 1979. *The Eighth Day of Creation*. New York: Simon and Schuster. Probably the definitive account of the discovery of the double helix.

Keller, E. F. 1983. *A Feeling for the Organism: The Life and Work of Barbara McClintock*. San Francisco: W. H. Freeman. A thorough account of Barbara McClintock's career.

Luria, S. E. 1973. *Life: The Unfinished Experiment*. New York: Scribner.

Luria, S. E. 1984. *A Slot Machine, a Broken Test Tube: An Autobiography*. New York: Harper and Row. Reminiscences of one of the original Phage Group. The slot machine refers to the insight that led to Luria and Delbrück's fluctuation test.

McCarty, Maclyn. 1985. *The Transforming Principle: Discovering that Genes Are Made of DNA*. New York: W. W. Norton. Personal recollections concerning Oswald T. Avery's studies on the transforming principle and the reactions of various members of the genetic community.

McElroy, W. D., and H. B. Glass, eds. 1957. *A Symposium on the Chemical Basis of Heredity*. Baltimore: Johns Hopkins University Press. One of the earliest symposia on the chemical nature of the gene.

Monod, Jacques. 1971. *Chance and Necessity*. New York: Knopf. Draws philosophical conclusions from allosteric proteins and DNA.

Monod, Jacques, and Ernest Borek, eds. 1971. *Of Microbes and Life*. New York: Columbia University Press. A collection of essays, most by participants in the search for the gene.

Moore, John A. 1985. *Science as a Way of Knowing*. Vol. 3: *Genetics*. American Society of Zoologists. One of an excellent series of publications designed to improve science education in high schools and colleges.

Morgan, T. H. 1926. *The Theory of the Gene*. New Haven, Conn.: Yale University Press.

Morgan, T. H., A. H. Sturtevant, H. J. Muller, and C. B. Bridges. 1915. *The Mechanism of Mendelian Heredity*. New York: Henry Holt.

Muller, H. J. 1946. A physicist stands amazed at genetics. Review of *What Is Life?* by E. Schrödinger. J. Hered. 37:90–92. A review that provides insight regarding an old-time geneticist's views of the book that was instrumental in converting many physicists into phage geneticists.

Olby, R. C. 1974. *The Path to the Double Helix*. London: Macmillan.

Peters, J. A., ed. 1959. *Classical Papers in Genetics*. Englewood Cliffs, N.J.: Prentice-Hall.

Ptashne, Mark. 1986. *A Genetic Switch: Gene Control and Phage λ*. Cambridge, Mass.: Cell Press. Details concerning the control of gene action at the best-studied locus.

Rosenfield, Israel, Edward Ziff, and Borin van Loon. 1983. *DNA for Beginners*. London: Writers and Readers.

Sayre, A. 1975. *Rosalind Franklin and DNA*. New York: W. W. Norton. Gives credit where credit is due.

Schrödinger, E. 1944. *What Is Life?* Cambridge: Cambridge University Press. A seminal book for many who later became outstanding molecular geneticists, but see H. J. Muller, 1946.

Sonneborn, T. M., ed. 1965. *The Control of Human Heredity and Evolution*. New York: Macmillan. Proceedings of perhaps the earliest conference in which participating authorities (Luria, Tatum, Pontecorvo, Muller, and others) were asked to address the impact the new genetics might have on human society.

Stadler, L. J. 1954. The gene. Science 120:811–819. An account of the search for the gene thirty-five years ago.

Stent, G. S. 1971. *Molecular Genetics: An Introductory Narrative*. San Francisco: W. H. Freeman. A textbook but written in narrative form.

Stern, Curt, and E. R. Sherwood, eds. 1966. *The Origin of Genetics. A Mendel Sourcebook*. San Francisco: W. H. Freeman.

Taylor, J. H., ed. 1965. *Selected Papers on Molecular Genetics*. New York: Academic Press.

Voeller, Bruce, ed. 1968. *The Chromosome Theory of Inheritance: Classic Papers in Development and Heredity*. New York: Appleton-Century-Crofts.

Weidel, Wolfhard. 1959. *Virus*. Ann Arbor: University of Michigan Press.

Zubay, G. L. 1968. *Papers in Biochemical Genetics*. New York: Holt, Rinehart, and Winston.

In addition to what may be regarded as standard texts, at frequent intervals the journal *Scientific American* publishes collections of reprints that deal with genetics; many of these—especially the earlier ones—have dealt with the search for the gene (dates at which the original papers appeared are listed in parentheses): *The Molecular Basis of Life: An Introduction to Molecular Biology* (1948–1968); *Facets of Genetics* (1948–1970); *Genetics* (1948–1981).

Index